J. Fehér · G. Csomós · A. Vereckei

Free Radical Reactions in Medicine

With 54 Figures

Springer-Verlag Berlin Heidelberg NewYork
London Paris Tokyo

Professor Dr. János Fehér
Semmelweis University of Medicine
Director, 2nd Department of Medicine
Szentkirályi u. 46
H-1088 Budapest

Professor Dr. Géza Csomós
Department of Clinical Research
Heimhuder Straße 46
D-2000 Hamburg 13

Dr. András Vereckei
VIII. Biró Lajos u. 17
H-1089 Budapest

Parts of this book were first published in:
János Fehér/András Vereckei: Szabadgyök-Reakciók Jelentösége Az Orvostudományban
by Medicina Könyvkiadó, Budapest, 1985

Library of Congress Cataloging in Publication Data.
Fehér J. (János), 1932-
Free radical reactions in medicine. Translated in part from: Szabadgyök-Reakciók Jelentösége Az Orvostudományban. 1985. Bibliography: p. 1. Free radicals (Chemistry)-Physiological effect. 2. Free radicals (Chemistry)-Toxicology. 3. Free radical reactions. I. Csomós, G. (GÉZA) 1935- II. Vereckei, A. (András), 1955- . III. Title [DNLM: 1. Free Radicals. QD 471 F296s]
QP527.F4413 1987 612'.01524 87-12704

ISBN-13: 978-3-642-83106-5 e-ISBN-13: 978-3-642-83104-1
DOI: 10.1007/978-3-642-83104-1

Softcover reprint of the hardcover 1st edition 1987

2121/3020-543210

Preface

Basic research has thrown new light on the importance of free radicals in biology and medicine. The clarification of interactions opens numerous physiological and pathological processes to closer scrutiny. The pathophysiological processes are now understood, even though the basic mechanisms of free radical reaction have as yet escaped full explanation.

A systematic approach makes it possible to re-evaluate and contrast the indications, reactions, and prospects for success.

The extracellular significance of oxygen radicals in inflammatory conditions and their role in the destruction of bacteria in the phagosome of granulocytes were primary factors intensifying research. Nature also provides a great number of examples which demonstrate that through the emergence of free radicals highly toxic combinations can form from innocuous substances. The reverse principle may also be encountered.

Taking into account free radical reactions in pathomorphological investigations helps prepare the ground for causal therapy. Research into these phenomena will contribute to a better understanding of physiological and pathophysiological processes.

Many questions still remain unanswered, and it will take several years of research before the final stone is unturned.

Budapest - Hamburg, June 1987

J. Fehér, G. Csomós, A. Vereckei

CONTENTS

Abbreviations

AA	arachidonic acid
ADP	adenosine diphosphate
ATP	adenosine triphosphate
BHA	Butylated Hydroxydnisole
C	complement
CAH	chronic active hepatitis
cAMP	cyclic adenosine monophosphate
CAT	catalase
CCA	lobenzarite disodium-N-(carboxyphenyl)-4-chloro-anthranylic acid disodium salt
cGMP	cyclic guanosine monophosphate
CH	cirrhotic hepatitis
DMSO	dimethyl sulphoxide
DPA	D-penicillamine
EMQ	ethoxyquin
GP	glutathione peroxidase
G-6-PD	glucose-6-phosphate dehydrogenase
GSH	reduced glutathione
HbS	haemoglobin S
HETE	hydroxyeicosatetraenic acid
$HO_2^- \cdot$	perhydroxy radical
H_2O_2	hydrogen peroxide
HPETE	hydroperoxyeicosatetraenic acid
IC	immune complex
LPO	lipid peroxidation
LT (A, B, C, D, E)	leukotriene (A, B, C, D, E)
MDA	malondialdehyde
MTDQ	(6,6-methylene-bis)-2,2,4-trimethyl-1,2-dihydroquinoline
MTDQ-DS	(2,2-dimethyl-4-methanesulphonic acid sodium-1,2-dihydroquinoline)
NADPH	reduced nicotinamide adenine dinucleotide phosphate
$O_2^- \cdot$	superoxide radical
$^1O_2$	singlet oxygen
$\cdot OH$	hydroxyl radical
PG	prostaglandin
PGI_2	prostacyclin
PMA	phorbol myristate acetate

PUFA	polyunsaturated fatty acid
R·	alkyl radical
RA	rheumatoid arthritis
RO·	alkoxy radical
ROI	reactive oxygen intermediate
ROO·	peroxy radical
ROOH	lipid hydroperoxide
SLE	systemic lupus erythematodes
SOD	superoxide dismutase
SRS-A	slow-reacting substance of anaphylaxis
TXA_2	thromboxane A_2

Introduction

The aim of the present work has been to provide insight into a new, rapidly developing field of medicine. Although the title may suggest a purely theoretical approach, our purpose has been to highlight the important practical aspects of the subject. Several detailed and comprehensive studies are available dealing with the chemistry and biology of free radical reactions; we, therefore, only wish to review the medical implications. It may seem premature to draw practical conclusions in a field where some of the basic biological and biochemical mechanisms have not yet been adequately clarified, still we feel that the potentialities involved justify our calling attention to them. This work is chiefly meant for doctors and medical students; therefore, after a chemical introduction, the role of free radical reactions in physiological processes and human diseases, and the relevant therapeutic possibilities will be discussed.

Recognition of the importance of free radical reactions in biology and medicine and the early results of research into their pathology are probably comparable to the description of DNA structure, expression of the genetic code, messenger function of cyclic adenosine monophosphate and the effect of prostaglandins.

In 1939 a protein was isolated from bovine erythrocytes, which was named erythrocuprein because of its copper content and localization. It was not until 30 years later that McCord and Fridovich discovered the enzyme nature of erythrocuprein and described the reaction it catalysed. With this discovery free radical pathology came into its own as a separate field of research. The practical implications of this new field are also important because we have effective tools for influencing biological free radical reactions. These are the antioxidants interfering with the reactions of free radical intermediates of oxygen. They occur naturally in the human body, but synthetic, nontoxic antioxidants are also available.

Free radicals of oxygen are formed in all aerobic organisms during both physiological and pathological processes. In the course of evolution, only those organisms survived which developed a system of defence against free radical reactions. Recent research has suggested that aerobic organisms possess a complex general defence system which might be compared to the immune system. Thus we may expect the recognition of a new group of disease entities called free radical diseases, similarly to the idea of autoimmune diseases - a category which was at first regarded with scepticism, but has by now become generally accepted. In the following, pathological and physiological processes will be dealt with, in the mechanism of which the role of free radical reactions has been proved or is highly probable.

Chapter I

The Chemistry of Free Radical Reactions

1 The Definition and Reactions of Free Radicals

The electron orbitals of most chemical substances contain paired electrons with opposite spins. Free radicals are molecules or fragments of molecules containing unpaired electrons in their outermost orbitals. Because of the natural tendency to electron pairing, the presence of these electrons results in a "tension", which induces the unpaired electron to "look for a partner". Therefore, most free radicals are extremely reactive and, as a result, short-lived. Some stable molecules, like nitric oxide NO and nitrogen dioxide NO_2, contain unpaired electrons in their outer orbitals and thus may be regarded as free radicals.

Biradicals are species containing two unpaired electrons, e.g. molecular oxygen. Radical ions are positively or negatively charged free radicals, e.g. H_3N^+, a protonated amine radical, or $O_2^-\cdot$, the superoxide radical anion. Whereas whether a species is an ion depends on the number of its electrons in relation to the number of protons, all species with unpaired electrons in their outer orbitals are free radicals, independent of the number of protons. Several chemical reactions are initiated by the fission of a two-electron bond. The bond may be cleaved symmetrically, producing free radicals (Eq. 1), or asymmetrically, when ions are formed (Eq. 2).

$$A:B \rightarrow A\cdot + B_+ \quad (1)$$
$$A:B \rightarrow A:^- + B^+ \quad (2)$$

Chemical substances may be converted to free radicals by irradiation (X-ray, UV), or by additives called initiators. It has long been known that ionizing radiations produce free radicals in biological systems, which play a major role in the aetiology of radiation injuries. However, it has been discovered only relatively recently that free radicals may be formed in aerobic organisms - including humans - in the course of physiological and pathological processes or by exogenous agents [112, 113, 443].

Free radical reactions are, in general, chain reactions. In the course of these, free radicals are formed during the initiation step (or steps). These enter into reactions in the course of several propagation steps, during which the number of free radicals is conserved until they eventually destroyed in a termination process. In the present work we deal with reactions of the free radicals and reactive intermediates of oxygen. As will be seen, the two are not identical because not all reactive intermediates of oxygen are free radicals.

2 Lipid Peroxidation

Lipids are readily oxidized both in vitro and in vivo, as shown for example by the peroxidized lipids present in age pigments. Molecular oxygen, which is a biradical, contains two unpaired electrons of parallel spin (spin is the quantized angular momentum of the electron which adds to the orbital angular momentum). The parallel spin prevents the direct addition of the electron pair to another molecule and thus chemical bond formation. For a bond to be formed spin reversal has to occur. Since, however, spin reversal takes a long time in comparison with the half-life of activated complexes[1], molecular oxygen is a relatively weak oxidant, and autooxidation of lipids in living organisms is a slow process. Nevertheless, if lipids are activated to free radicals R· via removal of hydrogen mediated by a free radical initiator, they enter into reaction with molecular oxygen more readily. A peroxy free radical ROO· is formed during the reaction, and this process is called lipid peroxidation (LPO). Polyunsaturated fatty acids (PUFA) are especially liable to peroxidation since the C–H bond of their so-called α-methylene carbon atom next to the double bond is weak, and thus they may be regarded as partially activated. Therefore, as the initial step of LPO, hydrogen is removed from this carbon atom (the hydrogen of the α-methylene carbon atom is called allylic hydrogen) [112, 443]. Figure 1 shows the process of LPO.

1. During initiation a free radical substance (X· in Fig. 1) reacts with a fatty acid chain containing one or more multiple bonds (R in Fig. 1, with two multiple bonds).

2. (a) A radical centre is formed on the carbon atom next to the double bond (α-methylene carbon) after the free radical initiator X· has abstracted a hydrogen and thereby ceased to be a free radical (XH). At the same time an alkyl radical R· is formed from the lipid. (b) Several configurational changes follow the initiation almost immediately - massive electron transfers occur: (i) transfer of the free electron to other C atoms; (ii) this moves the double bonds closer, leading to the formation of conjugated double bonds, which can be demonstrated by spectrophotometry as signs of early radical damage; (iii) some double bonds adopt the *trans* configuration, instead of the *cis* configuration characteristic of fatty acids.

3. The oxygen is added to the alkyl radical and a peroxy radical ROO· is formed.

4. Peroxy radicals abstract hydrogen from the nearby molecules (RH, which may be other unsaturated fatty acids, proteins, antioxidants, or nucleic acids) and thus metastable lipid hydroperoxides ROOH are formed; these products are transiently stable.

5-6. Hydroperoxide is capable of spontaneous decomposition to ·OH and alkoxy radicals RO·. As a result of metal catalysis alkoxy and peroxy radicals and hydroxyl or hydrogen ions are formed.

---

[1] Activated complexes are high-energy, loose, transient complexes formed in the course of reactions involving 2-3 molecules. They form and are degraded rapidly and are responsible for chemical reactions [142].

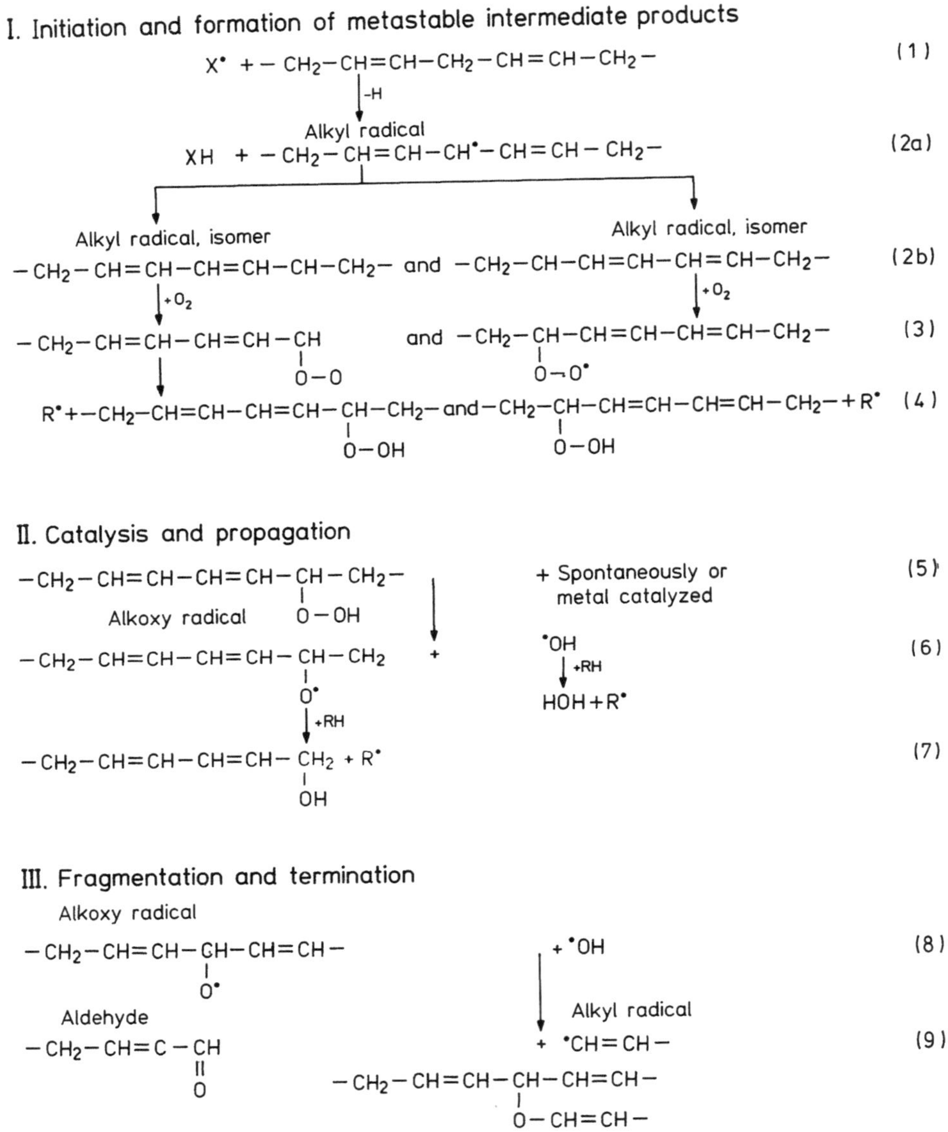

Fig. 1. Phases of lipid peroxidation, for details see text. (After Demopoulos et al. [115])

7. Owing to the abstraction of hydrogen from neighbouring molecules RH, the alkoxy and hydroxyl radicals are terminated, but further radicals are formed from the molecules losing their hydrogens.

8. Alkoxy radicals may undergo further oxidation or may be fragmented by continuous free radical attack on other *a*-methylene carbon atoms. In this way bis-hydroperoxides (i.e. two OOH groups on single fatty acid) may be

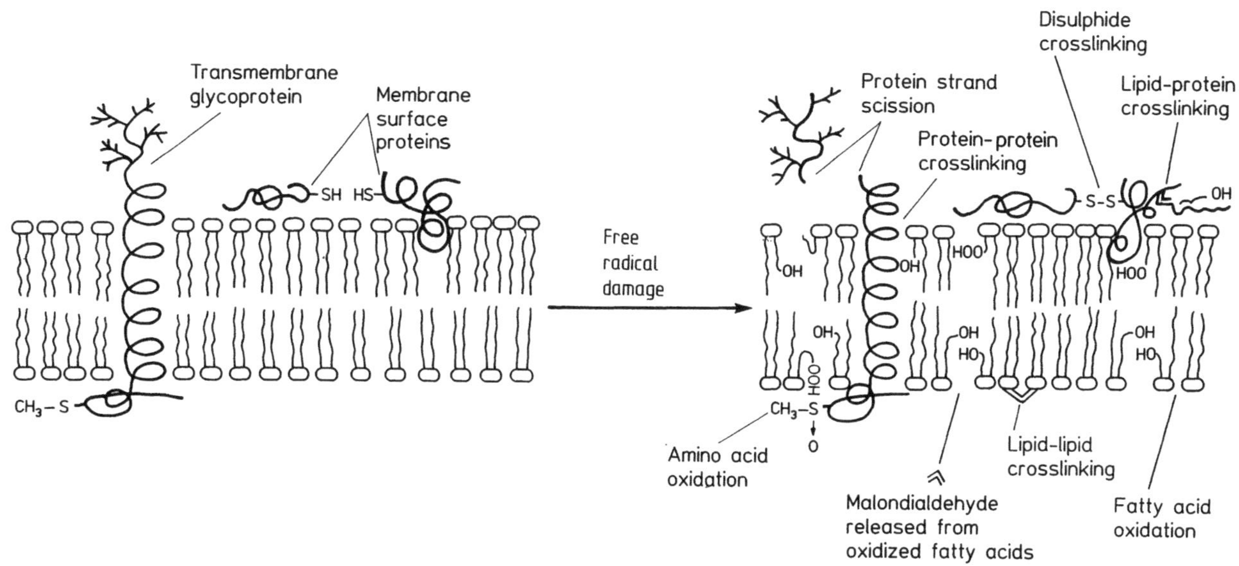

Fig. 2. Free radical induced membrane damage. Malondialdehyde, one of the products of lipid peroxidation, facilitates cross-linking. Free radicals may also produce protein-protein cross-links, amino acid oxidation a fissure of protein fibres. (After Freeman and Crapo [176])

formed and the fatty acid is decomposed into aldehydes and alkyl radical fragments.

9. Alkyl radicals may react with surrounding radicals, e.g. alkoxy radicals, and thus the process is terminated as a result of the formation of O-bonded bridges between the molecules, or C–C bonds may be formed with another alkyl radical.

Initiation may thus start a chain reaction which, if uncontrolled by defence mechanisms, may lead to extensive damage of the surrounding molecules. Autooxidation of lipids eventually leads to the formation of new C–C bonds. Since lipids are membrane constituents, it is easy to see to what extent cross-linking may damage the membranes (Fig. 2).

Autooxidation is considerably accelerated by the presence of transition metals (iron, copper) enhancing the degradation of metastable hydroperoxides (Fig. 1. Eqs. 5, 6). Lipids, which are readily oxidized, promote the metal-catalysed oxidation, in the presence of O_2, of more inert cellular components like proteins or nucleic acids (Fig. 1. Eqs. 7, 9). Autooxidation is a very slow, circumscribed process under physiological conditions. Under pathological conditions, however, it is accelerated, becomes more extensive and may damage all basic biomolecules [115, 443, 503]. The metastable lipid peroxides formed during LPO, malondialdehyde (MDA), and other stable end products of LPO, e.g. 2-alkenals and 4-hydroxyl-2-alkenals, may reach more distant parts of the cell or other cells and tissues owing to their relative stability. Thus LPO may also damage cells or tissues which are not directly exposed to peroxidative damage [431, 446].

3 The Most Important Physical and Chemical Properties of Molecular Oxygen: Radicals of Molecular Oxygen

Radicals formed from molecular oxygen are a serious hazard for both living and nonliving matter. Free radical reactions in the environment occur, e.g. (a) when oxygen reacts with hydrocarbons in car engines; (b) when smog is formed, in which the hydroxyl radical, generated when nitric acid is irradiated at wavelengths less than 400 nm, plays a major role. Environmental free radicals may cause the corrosion of metals and plastics and play a role in the rancidification of butter and fats [229, 351].

As mentioned, molecular oxygen may be regarded as a biradical. This property allows it to act as an initiator and to participate in additive reactions. Nevertheless, it is basically a relatively weak oxidant owing to its free electrons with parallel spins. Reactive intermediates of oxygen are generated by reduction or excitation (Fig. 3). As shown in the figure, the reactive intermediates formed in the course of excitation and reduction are not invariably free radicals (e.g. delta singlet oxygen or H_2O_2 are not free radicals themselves, nevertheless, since they resemble free radicals in their reactivity and importance. Therefore, they will be discussed together with free radicals under the heading “reactive oxygen intermediates (ROIs)“.

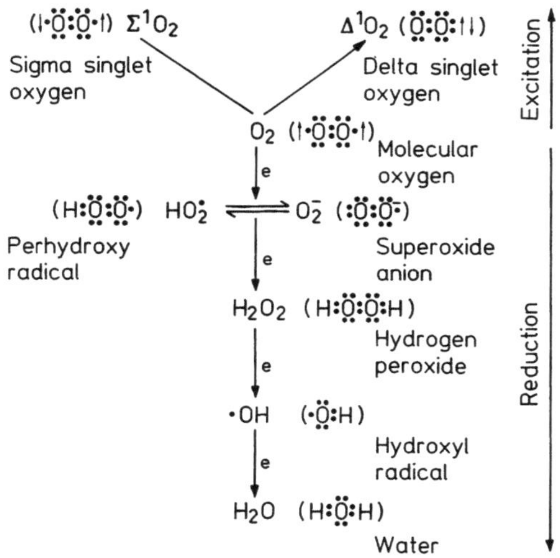

Fig. 3. Reactive oxygen intermediates produced during reduction an excitation of oxygen. (After Klebanoff [302])

3.1 Reduction

In the course of the complete tetravalent (four-electron) reduction of oxygen, water is formed. If it undergoes sequential univalent (one-electron) reduction, reactive intermediates are formed, which may damage biological systems. The superoxide radical anion $O_2^{-}\cdot$, hydrogen peroxide and the hydroxyl radical ·OH are such intermediates. According to the quantum mechanical concept of spin restriction,[2] univalent reduction is more likely to occur in vivo. In living organisms, therefore, oxidative enzymes have developed, which circumvent the law of spin restriction and catalyse divalent and tetravalent reduction of oxygen, so that only small amounts of ROIs are formed. Cytochrome oxidase belongs to these enzymes. It catalyses the tetravalent reduction of oxygen and it is responsible for the bulk of total oxygen consumption in the course of cellular metabolism.

Nevertheless, ROIs are formed in small amounts during cellular metabolism, but the defence systems are able to neutralize them under normal conditions. Further ROIs are formed during autooxidation of different chemical compounds, most easily via univalent reduction of O_2 (e.g. at autooxidation of hydroquinones, leukoflavins, catecholamines, thiols, tetrahydropterins, or oxyhaemoglobin, $O_2^{-}\cdot$ is formed). In addition, ROIs are generated during enzymatic processes (e.g. xanthine oxidase), in subcellular organelles (mitochondrium, chloroplast), or during cell functions (of phagocytes or thrombocytes).

[2] The Pauli principle states that an atom cannot contain two electrons with four identical quantum numbers.

3.2 Excitation

Singlet oxygen $^1O_2$ is formed if one of the unpaired electrons of molecular oxygen is transferred via energy absorption to a higher energy orbital and its spin is inverted. Singlet oxygen exists in two states, i.e. as delta singlet oxygen Δ^1O_2 in which the newly paired electrons occupy the same orbital, and sigma singlet oxygen Σ^1O_2 in which the electrons of opposite spins are in different orbitals. The surplus energy of singlet oxygen may be spent through thermal decomposition, light emission, or chemical reaction (as indicated by chemiluminescence) [179, 302, 360].

4 Reactions in which ROIs are Formed

$O_2^-\cdot$ may be formed, as has been mentioned, during autooxidation of various compounds, in the mitochondrial respiratory pathway, if the members of the cycle are reduced and react with oxygen, in chloroplasts, in connection with phagocyte function and NADPH oxidase function (see Chap. III Sects. 1, 3). The reactivity of $O_2^-\cdot$ is limited, its importance lies in its ability to oxidize transition metal complexes and organic substrates, to bind to metals, and to be transformed into other, more reactive oxygen intermediates. In acid media, such as the vacuole of the phagocyte or the microenvironment of cell membranes (the cell surface, containing polyanions, attracts H^+ ions), $O_2^-\cdot$ is protonated into the perhydroxy radical $HO_2^-\cdot$

$$H^+ + O_2^-\cdot \rightarrow HO_2^-\cdot$$

$HO_2^-\cdot$ is a stronger oxidant than $O_2^-\cdot$ and is cytotoxic. $\cdot OH$ formation is another important reaction of $O_2^-\cdot$ (Haber-Weiss reaction, Fenton reaction) [176, 177].

H_2O_2 is formed either by the divalent reduction of molecular oxygen catalyzed for example by glucose oxidase or from $O_2^-\cdot$ catalyzed by the enzyme superoxide dismutase (SOD) (see Chap. II Sect. 4).

A role is attributed in the formation of the extremely reactive radical $\cdot OH$ to the Haber-Weiss reaction, in the course of which $\cdot OH$ is produced from $O_2^-\cdot$ and H_2O_2

$$O_2^-\cdot + H_2O_2 \rightarrow O_2 + OH^- + \cdot OH$$

This reaction, however, proceeds very slowly in living organisms and thus no significant amounts of $\cdot OH$ are formed. The reaction is accelerated by metal catalysis, resulting in a Fenton-type reaction

$$\begin{array}{lcl} O_2^-\cdot + M^{n+1} & \rightarrow & M^n + O_2 \\ M^n + H_2O_2 & \rightarrow & M^{n+1} + \cdot OH + OH^- \\ \hline O_2^-\cdot + H_2O_2 & \rightarrow & O_2 + \cdot OH + OH^- \end{array}$$

Significant amounts of $\cdot OH$ may thus be produced. By analogy with the Fenton reaction hydroperoxide may participate in the reaction instead of H_2O_2

$$Fe^{2+} + ROOH \rightarrow Fe^{3+} + RO\cdot + OH^-$$

The alkoxy radical $RO\cdot$ is only slightly less reactive than $\cdot OH$, and may damage biological systems excessively either itself or as a hydroxyalkyl radical produced as a result of internal hydrogen atom transfer.

Singlet oxygen may be formed in the following reactions:

$$HO_2^-\cdot + O_2^-\cdot + H^+ \longrightarrow {}^1O_2 + H_2O_2 \qquad (1)$$

$$O_2^-\cdot + H_2O_2 \xrightarrow{\text{metal}} {}^1O_2 + OH^- + \cdot OH \qquad (2)$$

$$OCl^- + H_2O_2 \longrightarrow {}^1O_2 + Cl^- + H_2O \qquad (3)$$

$$O_2^-\cdot + \cdot OH \longrightarrow {}^1O_2 + OH^- \qquad (4)$$

$$2\,O_2^- + R{-}\overset{\displaystyle O}{\overset{\|}{C}}{-}O{-}O{-}\overset{\displaystyle O}{\overset{\|}{C}}{-}R \rightarrow 2\,{}^1O_2 + 2\,RCO_2^-\cdot \qquad (5)$$

Reaction 1 is an analogue of the spontaneous dismutation of $O_2^-\cdot$ (spontaneous dismutation is the same as the SOD-catalysed reaction, but is a nonenzymatic reaction and is much slower.) Reaction 2 is a variant of the Fenton reaction. Reaction 3 occurs in the course of the function of the granulocyte myeloperoxidase-H_2O_2-halide system (see Chap. III Sect. 1). ${}^1O_2$ may be formed during the interaction of $O_2^-\cdot$ and $\cdot OH$ (reaction 4) and of $O_2^-\cdot$ and diacyl peroxides (reaction 5) [18, 147].

5 Summary

Free radicals are molecules or fragments of molecules containing unpaired electrons in their outer orbitals, which represent a free valence. Unpaired electrons tend to acquire a pair, therefore, free radicals are chemically highly reactive and have a short half-life. Free radical reactions are nonenzymatic chain reactions. Free radicals of oxygen play an important role in biology and medicine. They are generated from oxygen by excitation (singlet oxygen ${}^1O_2$) or reduction (*superoxide anion radical* $O_2^-\cdot$, *hydrogen peroxide* H_2O_2, *hydroxyl radical* $\cdot OH$). *Not all reactive products of the reduction or excitation of oxygen are free radicals (e.g. singlet oxygen, hydrogen peroxide), but their reactivity and reactions resemble those of free radicals. These products, together with the true free radicals are called reactive oxygen intermediates* (ROIs). *In living organisms free radicals may be formed by exogenous interactions (e.g. UV, X-rays chemicals), but also during physiological processes. Under physiological conditions free radicals may be formed in all parts of the cell (e.g. in the mitochondrial and microsomal electron transport chain, in the course of autooxidation of biomolecules, cytoplasmic enzyme function, or in the plasma membrane of phagocytes). Physiological free radical reactions are under strict control, but the failure of these mechanisms may give rise to pathological free radical reactions. Autooxidation of the lipids of living organisms is a slow, circumscribed process because molecular oxygen is a weak oxidant. If, however, a free radical extracts a hydrogen from the lipid* R *producing a lipid free radical* R·, *this will more readily react with molecular oxygen, leading to the formation of a peroxy free radical* ROO·. *This process is called lipid peroxidation (LPO), which is only the beginning of a chain reaction.* ROIs *formed during the process of LPO can damage also nonlipid biomolecules (proteins, nucleic acids, carbohydrates). Iron and copper catalyse and thus accelerate LPO.*

Chapter II

Control of Free Radical Reactions in Biological Systems

Since living organisms contain several free radicals under physiological conditions, there must exist innate mechanisms keeping their reactions under control in situations like electron transport, melanin production or photosynthesis. Molecular structures of vital importance like the cell membrane. DNA and proteins have constituents which are very sensitive to radical reaction, such as the α-methylene carbon atom of PUFAs as well as allylic hydrogen, and the thymidine base.

1 Control by Membrane Structure

Free radicals formed during electron transport are kept under control by the enzymes and carriers of the transport chain, which are bound to the membrane and are closely packed together. The PUFAs of the membrane are localized in its hydrophobic region, far from the site of electron transport, which takes place in the hydrophilic region. Thus the protection of PUFAs is inherent in membrane structure. This defence may be damaged if the cell membrane or mitochondria are injured, giving rise to pathological free radical reactions. PUFAs of the membrane phospholipids are "packaged" by cholesterol. Hydrogen bonds in membranes also provide protection against free radicals. However, hydrophobic chemical pathogens (e.g. CCl_4, DDT, halothane and some carcinogenic substances) can penetrate the hydrophobic region and initiate free radical reactions.

2 Control by Specific Molecular Structure

Some normally occurring free radicals are under control because they are intermediates formed during reactions of enzyme-substrate complexes. Owing to their fixed localization and short half-life, they have little chance to interact with other substances.

The rate of radical reactions is influenced by the amount of membrane unsaturated fatty acids and the degree of their unsaturation. These, in turn, depend to a lesser extent on the fat content of the diet, and predominantly on genetic factors [114].

3 Scavengers and Antioxidants

Substances interacting with free radicals and thereby inhibiting free radical reactions are called scavengers. These must reach their target at the right time and in the right concentration. Synthetic scavengers used for therapy must be of low toxicity. Also, the scavenger radicals formed during the interaction of the scavenger with the toxic radical intermediate should be less reactive (i.e. have a longer half-life) than the radical they attack. Scavengers show a high degree of selectivity in their reactions with free radicals, which are stoichiometric reactions. They should therefore be applied in proper doses [504].

Substances which provide protection against the toxic effects of ROIs are called antioxidants. Only "natural" antioxidants occurring in biological systems under physiological conditions will be dealt with in this chapter. These include the vitamins C, A, E, K, selenium, thiol-containing compounds like cysteine, cysteamine, glutathione, methionine, gallic acid derivatives, ubiquinone, urate and glucose.

Antioxidants may attack in different phases of LPO. They may: (a) inhibit the initiation process (abstracting the allylic hydrogen from the α-methylene carbon atom), e.g. vitamin E; (b) inhibit the formation of hydroperoxides (chain breakers like vitamins E and C); (c) degrade the hydroperoxides formed without producing radicals (e.g. glutathione peroxidase, thiols); (d) act as metal chelating substances (inhibiting the metal catalysis accelerating the decomposition of hydroperoxides), e.g. D-penicillamine; (e) remove the free radical (scavenger activity), e.g. vitamins E and A [25, 263].

Individual antioxidants may act by more than one of these mechanisms, e.g. the synthetic antioxidants butylated hydroxyanisole (BHA), butyl hydroxytoluene (BHT), ethoxyquin (EMQ), and the dihydroquinoline derivative MTDQ-DS inhibit initiation, interrupt the reaction chain, destroy hydroperoxides, and also have a scavenger action, i.e. they interfere with all phases of LPO [25]. Whether a substrate behaves as an antioxidant or prooxidant (facilitating peroxidation) depends on its concentration and on the environment in which it acts. As an example, let us suppose that free radical X· is formed in a cell and attacks cell constituent A in order to abstract an electron to be paired with its single electron. This leads to further damaging reactions. Then antioxidant Z with a scavenger action appears, captures the unpaired electron of X· and reacts itself. The same scavenger molecule Z may, under different conditions, be activated to Z· after having accepted the single electron of X· and may be even more reactive than X·, causing even greater damage. Therefore, if using synthetic scavengers, it is an important requirement that the scavenger radical product of the reaction should be less reactive than the radical to be removed. At the same time, reactions in complex in vivo systems are not completely predictable. For example, copper, if added to PUFA, facilitates autooxidation. However, copper complexes in the tissues generally (but not invariably) behave as antioxidants. Vitamin C might be quoted as another example. It is a prooxidant in low concentrations in the presence of trace amounts of transition metals (iron, copper) - either free or chelates - because it reduces the metals, thereby promoting the metal catalysis of LPO. In high concentrations, however, when there is a sufficient number of metal binding sites, vitamin C is an antioxidant.

Vitamin C is a strong reducing agent. It is unstable, readily losing its hydrogen atoms. It may act as an antioxidant by interrupting the chain reaction: losing a hydrogen, it reacts with the peroxy radicals and stable monodehydroascorbate is produced. In addition, it may also have a direct $O_2^-\cdot$ and $\cdot OH$ scavenger action. When vitamin C interacts with ROIs, the product is a mixture of monodehydroascorbate and dehydroascorbinic acid (depending on whether one or two hydrogen atoms are removed) [126, 439].

The main lipid phase antioxidant activity of vitamin E in the cell membrane occurs via hydrogen donation to hydroperoxides; thus the formation of hydroperoxides is prevented, the chain reaction is terminated, and the extension of the pathological free radical reaction is prevented. In addition, it also scavenges $^1O_2$. The phytyl side chain of vitamin E is not involved in its antioxidant action, its role is to get the chroman side chain responsible for the antioxidant action into the optimum position [263].

Vitamin A plays a role in the differentiation of epithelial tissue and in growth, and has anticarcinogenic action. Its anticarcinogenic and antioxidant actions may be in connection with its unsaturated fatty acid content, which provides an alternative pathway for LPO. It is also a $^1O_2$ scavenger, and its metal chelating property contributes to its antioxidant and anticarcinogenic actions.

The antioxidant property of vitamin K is due to its quinone structure, capable of redox reactions.

Ubiquinone (coenzyme Q) has a structure resembling those of vitamin E and vitamin K, and is a constituent of the mitochondrial respiratory chain. It is an antioxidant; for example, the symptoms of vitamin E deficiency in animals showed improvement in response to the administration of coenzyme Q.

Selenium is a constituent of the enzyme selenium-dependent glutathione peroxidase (SeGP), which prevents LPO. It also has a membrane stabilizing action.

Thiols interfere with free radical reactions in general, by abstraction of the hydrogen of the -SH group. The "thiyl" radical -S· produced may interact with another, similar radical to form disulphide -S-S- [25, 125, 171, 263,

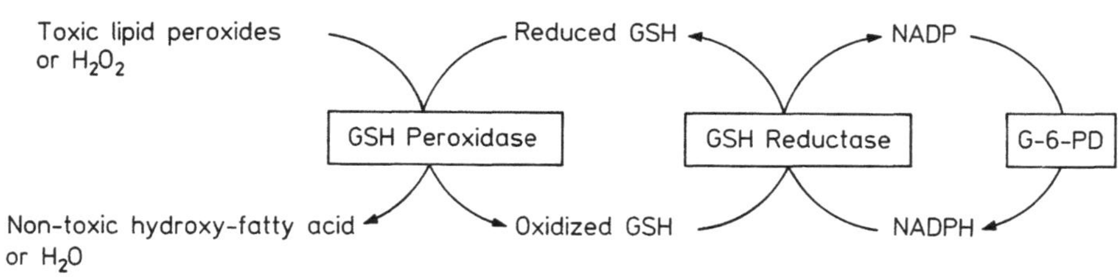

Fig. 4. The glutathione peroxidase system. GSH reduced glutathione; G-6-PD glucose-6-phosphate dehydrogenase; NADP nicotinamide adenine dinucleotide phosphate (oxidized form); NADPH nicotinamide adenine dinucleotide phosphate (reduced form). (After Frank and Massaro [175])

612]. The function of the SeGP-thiol system found in the cytosol and in mitochondria and utilizing reduced glutathione is shown in Fig. 4. A direct scavenger action of reduced glutathione has also been suggested [175]. Antioxidant action of urate has been reported [176, 263, 507], which is probably due to a direct ·OH and $^1O_1$ scavenger mechanism.

Glucose has a weak ·OH scavenger action. Since it is present in the extracellular space in high concentration under physiological conditions, its effect may be considerable [477].

Antioxidants may interact with one another, e.g. vitamin C may regenerate vitamin E by donating a hydrogen atom to the vitamin E radical formed after hydrogen transfer from vitamin E to the peroxy radical [400].

4 Cellular Enzymatic Defence Against Radicals of Molecular Oxygen

It must be emphasized in advance that the chief mechanism protecting cells in vivo against oxygen toxicity and LPO is the maintenance of low oxygen tension in the tissues (approximately 26 mm Hg). Thus the formation of $O_2^-\cdot$ and other LPO-inducing ROIs is kept under control [263].

Removal of $O_2^-\cdot$ is mediated by the SOD enzymes.

5. In recent years unexpected progress has been made in the sphere of research into superoxide dismutases (SOD). It has taken almost 50 years before observation evolved into a definition and affiliation to a biological system. In 1939 Mann and Keilin isolated for the first time a protein containing copper from erythrocyte, which came to be named haemocuprein. In the coming years different species of more cupreins were isolated and their identity compared to one another confirmed. In 1965 Huber isolated a type of protein from cattle liver which contained copper and zinc in identical quantity.

In 1969 McCord and Fridovich discovered the enzymatic activity of erythrocuprein. They recognized that the erythrocuprein changes superoxide radical into molecular oxygen and H_2O_2 and came to call it superoxide dismutase.

This discovery led to an intensive examination of the biological significance of radical oxygen. In 1974 Steinmann fully succeeded in clarifying the structure. The superoxide dismutase is a metallic protein consisting of two identical subunits. All subunits are composed of 151 amino acids and contain an atom of copper and zinc each:

SH (above Cy) 10 (above Gly)
Ac-Ala-Thr-Lys-Ala-Val-Cy-Val-Leu-Lys-Gly-Asp-Gly-Pro-Val-Gln-
20 (above Phe) 30 (above Thr)
Gly-Thr-Ile-His-Phe-Glu-Ala-Lys-Gly-Asp-Thr-Val-Val-Val-Thr-Gly-Ser-Ile-
40 (above Asp) 50 (above Asp)
Thr-Gly-Leu-Thr-Glu-Gly-Asp-His-Gly-Phe-His-Val-His-Gln-Phe-Gly-Asp-
60 (above Pro)
Asn-Thr-Gln-Gly-Cy-Thr-Ser-Ala-Gly-Pro-His-Phe-Asn-Pro-Leu-Ser-Lys-
70 (above Gly) 80 (above Gly)
Lys-His-Gly-Gly-Pro-Lys-Asp-Glu-Glu-Arg-His-Val-Gly-Asp-Leu-Gly-Asn-

90 100
Val-Thr-Ala-Asp-Lys-Asn-Gly-Val-Ala-Ile-Val-Asp-Ile-Val-Asp-Pro-Leu-Ile-
110
Ser-Leu-Ser-Gly-Glu-Tyr-Ser-Ile-Ile-Gly-Arg-Thr-Met-Val-Val-His-Glu-
120 130
Lys-Pro-Asp-Asp-Leu-Gly-Arg-Gly-Gly-Asn-Glu-Glu-Ser-Thr-Lys-Thr-Gly-
140 150
Asn - Ala - Gly - Ser - Arg - Leu - Ala - Cy - Gly - Val - Ile - Gly - Ile - Ala - Lys

Both subunits are bound by a disulphide bridge, and lose their enzyme activity through the dissolution. The molecule has a bullet-shaped spatial structure and the external copper atoms make up the active catalytical centre. Zinc plays no role at all in the enzyme activity.

The therapeutic pontential of superoxide dismutases (SOD) has stimulated various companies to produce an authentic human SOD (HSOD) by gene technology.

These are metalloprotein containing Mn, Fe or Cu + Zn (MnSOD, FeSOD, CuZnSOD) and are ubiquitous in the cells of aerobic organisms. FeSOD is found in prokaryotes, MnSOD in both prokaryotes and eukaryotes, while CuZnSOD is found only in eukaryotes. The mitochondria of eukaryotes contain only MnSOD, whereas the cytosol also contains MnSOD in addition to CuZnSOD. SOD catalyses the reaction

$$2\,O_2^{-}\cdot + 2\,H^+ \rightarrow H_2O_2 + O_2$$

H_2O_2 may thus be formed not only during the divalent reduction of O_2, but also by SOD activity. For instance, MnSOD may transform $O_2^{-}\cdot$ produced during mitochondrial metabolism into H_2O_2, which means that there is a permanent production of small amounts of H_2O_2 in the cells.

H_2O_2 is removed by catalase (CAT), a heme enzyme, and by various peroxidases (e.g. SeGP). CAT is found in peroxisomes and catalyses the reaction

$$H_2O_2 \rightarrow H_2O + \frac{1}{2}O_2$$

At low intracellular H_2O_2 concentrations CAT is relatively unreactive. Its peroxisomal localization also suggests that other defence systems may also be responsible for the removal of H_2O_2. Thus peroxidases catalyse the reaction

$$H_2O_2 + RH_2 \rightarrow 2\,H_2O + R$$

They decompose H_2O_2 utilizing reducing agents. SeGP can remove not only H_2O_2, but also lipid hydroperoxides. Thus it functions later than vitamin E: it does not prevent the formation of hydroperoxides, but degrades them once they have been formed. At low H_2O_2 concentration SeGP provides the main line of defence, whereas at higher, toxic levels of H_2O_2 CAT also starts functioning. SOD and peroxidase or SOD and CAT may jointly fulfil the role of a terminal oxidase, catalysing the reduction of $O_2^{-}\cdot$ to water [178, 179, 351, 360]. Recently a selenium-independent glutathione peroxidase (SeiGP) has been isolated, which cannot metabolize H_2O_2, only organic peroxides. The role of SeGP is probably in protection against LPO [261].

The cellular enzymes described - with the exception of SeGP and SeiGP, which attack lipid peroxides - remove ROIs directly, thereby inhibiting LPO, which is kept at a minimum level. The cellular antioxidants described in Sect. 3 constitute a second line of defence in addition to their participation in the first via their scavenger activity.

There is a third line of cellular defence, too, decomposing the end products of LPO, the alkenes, alcohols, epoxides and aldehydes produced in the degradation of lipid peroxides. It has been shown that enzymes that have so far been thought to have the primary function of metabolizing xenobiotics, have been shown to be capable of metabolizing LPO products. Possibly this activity also belongs to their primary functions. For instance, epoxide hydrolase and aldehyde reductase have been shown to metabolize peroxidation products, while cytochrome P-450 monooxidases have been found to metabolize endogenous lipids and prostaglandins. Cytochrome P-450 has multiple functions. It may participate in defence mechanisms against LPO, may be the source of $O_2^-\cdot$ formation, and may itself be damaged during LPO (see Chap. III Sect. 3) [263]. The protective enzymes are mutually interdependent. For instance, $O_2^-\cdot$ inhibits CAT and other heme-containing peroxidases, and SOD provides protection against this inhibitory effect. Two of the three SODs are inactivated by H_2O_2, while CAT and peroxidase provide protection against this effect [177].

5 Antioxidant Defence of the Extracellular Space

Only a very low level of the enzymes participating in cellular defence can be detected in the extracellular space. They probably only get there by passive difusion from the interior of the cells. Therefore, the defence of the extracellular space against free radical reactions is much weaker. In addition, only a small fraction of the low serum SOD activity is due to CuZnSOD, its greatest part is accounted for by a high molecular weight tetrameric SOD (CuZnSOD is a dimer), which also contains copper (four atoms) and probably the same amount of zinc. It was believed earlier that this tetrameric SOD only occurs in the extracellular space, as indicated by its glycoprotein structure characteristic of most plasma proteins, but not found intracellularly. It was, therefore, called extracellular SOD (EC-SOD). Recently, however, EC-SOD has been detected in cells, too. Its exact localization is not known, and, apart from a few exceptions, it occurs in much smaller amounts intracellularly than CuZnSOD and MnSOD. The SOD activity in the extracellular space is low in spite of the presence of EC-SOD. This probably reflects a “compromise” between the necessary defence of the extracellular space against destructive ROIs and the protection of the useful functions of ROIs (e.g. the microbicidal and cytolytic function of phagocytes and the chemotactic activity triggered by $O_2^-\cdot$ [341, 349, 350]. Coeruloplasmin has not only a copper transporting function, but is also a ferroxidase; it oxidizes Fe(II) to Fe(III), thereby inhibiting the metal catalysis of LPO. It probably also acts as a direct scavenger of $O_2^-\cdot$. These properties make coeruloplasmin one of the antioxidants of extracellular space.

Transferrin is believed to be another circulatory antioxidant. As opposed to coeruloplasmin, which is invariably saturated with copper, transferrin is saturated with iron only in pathological cases. Its antioxidant activity depends on its iron-free fraction only and is due to iron chelation. An antioxidant action has been attributed to urate and glucose in the extracellular space [8, 125, 147, 263, 477].

6 Summary

Living organisms are protected against the potential damaging effects of physiological free radical reactions by strict control mechanisms. Certain characteristic features of membranes are responsible for their partial protection against pathological free radical reactions. The molecular environment strongly influences free radical reactions. Antioxidants are substances providing defence against the toxic effects of ROIs. Natural antioxidants (vitamins A, C, E, K, selenium, coenzyme Q, thiol-containing compounds) and synthetic ones are different. Antioxidants interfere with LPO in its different phases. Those removing the free radicals by reacting with them are called scavengers. Prooxidants are substances facilitating LPO. The same substance may behave as a prooxidant or an antioxidant depending on its concentration and on the molecular environment. The main mechanism of in vivo defence against oxygen toxicity is the maintenance of low oxygen tension. The cells contain various enzymes for the removal of ROIs. The superoxide radical anion $O_2^{-}\cdot$ *is removed by superoxide dismutase (SOD), while hydroperoxide* H_2O_2 *is removed by CAT and glutathione peroxidase. These constitute the primary line of cellular defence, inhibiting the initiation of LPO. Antioxidants occurring naturally in cells constitute the second line of defence, although as scavengers they also participate in primary defence. There is also a third line: decomposition of the endproducts of LPO mediated probably by enzymes of xenobiotic metabolism, e.g. epoxide hydrolase and cytochrome P-450. Compared with the cells, the extracellular space is less efficiently protected against free radical reactions. A high molecular weight SOD, coeruloplasmin, together with transferrin and probably also glucose and urate are responsible for its antioxidant protection.*

Chapter III

Physiological Free Radical Reactions

Physiological free radical reactions occur under controlled conditions and play an important role in the function of various cells, subcellular organelles and enzymes, and in the synthesis and degradation of biomolecules. ROIs may be formed in all parts of the cell (Fig. 5).

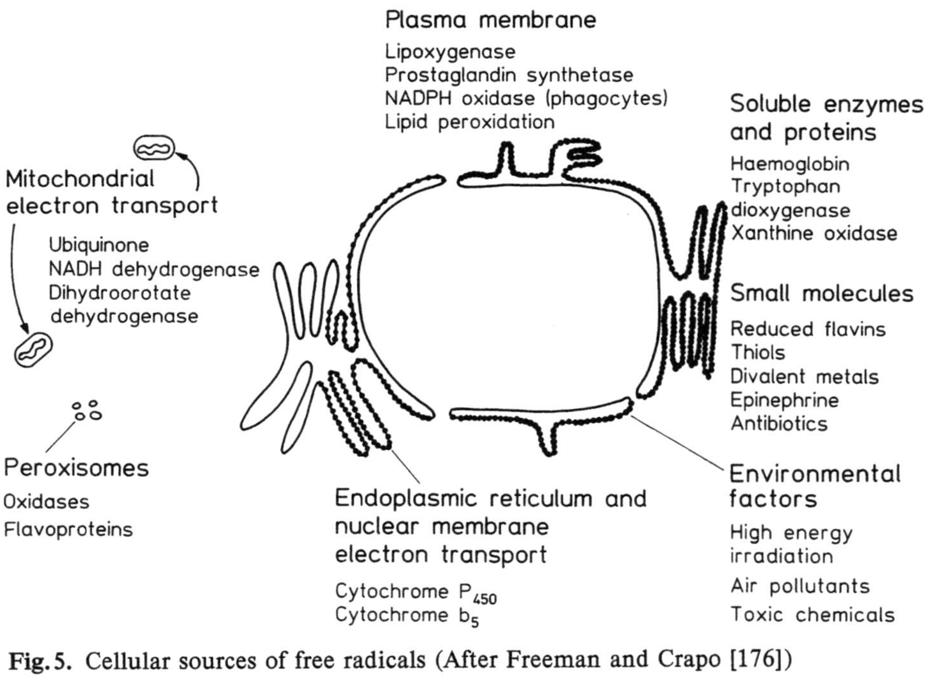

Fig. 5. Cellular sources of free radicals (After Freeman and Crapo [176])

1 The Role of Free Radical Reactions in Phagocytosis

If phagocytes (neutrophils, eosinophils, mononuclear phagocytes) are exposed to an appropriate stimulus, a coordinated series of reactions occurs in them, which is called a respiratory burst. Both particulate and soluble agents are capable of activating the phagocytes. Particulate agents include for example opsonized bacteria, zymosan derived from the cell wall of yeast, and latex spherules, which do not need opsonization. Phorbol myristate acetate (PMA), lymphokines, C5a, immune complexes, and immunoglobulins are soluble activators. The respiratory burst provides oxidizing agents for the destruction of the target cells by phagocytosis. Its activation is a complex event started by the interaction of the stimulus with the plasma membrane of the phagocyte. Phagocytosis or degranulation are not preconditions of activation either by soluble or by particulate agents. The respiratory burst may be activated even in the absence of these processes. However, energy is needed for its activation. The molecular mechanism of this activation process is insufficiently known; a role has been attributed to Ca^{2+} redistribution (to the Ca^{2+}-dependent enzymes of cytosol), to the unsaturated fatty acids (e.g. arachidonic acid) formed as a result of phospholipase A and C activation, to lysophospholipids (phospholipids from which a fatty acid has been split off under the catalytic effect of phospholipase A), and to diacylglycerol (derived from phospholipids by removal of a polar phosphoric acid part catalysed by phospholipase C). The stimulus triggers the following process

$$\text{stimulus} \xrightarrow{10\,s} \text{membrane depolarization} \xrightarrow{30\text{–}60\,s} O_2^{-}\cdot \text{ formation, change in cAMP level, lysosomal enzyme release/degranulation}$$

$O_2^{-}\cdot$ formation concomitant with degranulation and a change in the intracellular cAMP level is thus preceded by membrane depolarization [184]. $O_2^{-}\cdot$ production starts as a result of the activation of membrane NADPH or NADH oxidase owing to membrane perturbation

$$2\,O_2 + NAD(P)H \rightarrow 2\,O_2^{-}\cdot + NAD(P)^{+} + H^{+}$$

It is not known whether NADH or NADPH is the physiological electron donor; according to most authors NADPH is the more likely one. The enzyme is localized in the membrane so that its NADPH binding site faces the cytoplasm where NADPH is found. The $O_2^{-}\cdot$-producing part is localized in the lipid bilayer of the membrane. This is why $O_2^{-}\cdot$ may be produced within the plasma membrane or on its outer surface. However, during phagocytosis the outer surface of the membrane faces the interior of the phagocytic vacuole, where $O_2^{-}\cdot$ is produced. It has been postulated that the enzyme transfers the

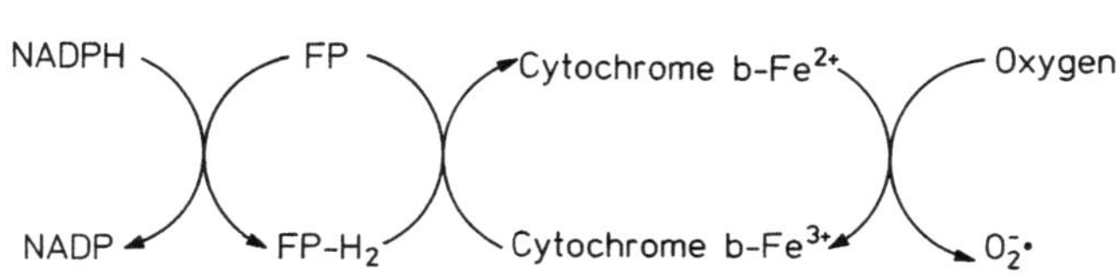

Fig. 6. Postulated $O_2^{-}\cdot$ -producing mechanism of the NADPH oxidase of phagocytes. FP flavoprotein. (After Babior [15])

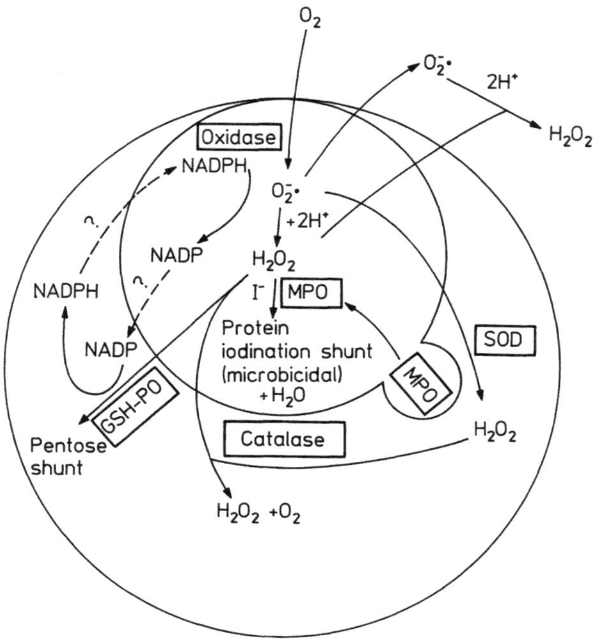

Fig. 7. Postulated pathway of the production and use of superoxide anion and hydrogen peroxide during phagocytosis. MPO myeloperoxidase; GSH-PO glutathione peroxidase; SOD superoxide dismutase. (After Root and Metcalf [467])

electron from NADPH to the oxygen via a mini electron transport chain (Fig. 6), in which a flavoprotein and cytochrome participate.

SOD transforms $O_2^-\cdot$ to H_2O_2 by dismutation (Fig. 7). The phagocytized particle or microorganism is contained by the vacuole, which fuses within the cell with the azurophilic or primary granules found in the neutrophilic and mononuclear phagocytes. The enzyme myeloperoxidase (MPO) of the azurophilic granules is a constituent of the H_2O_2-MPO-halide system utilizing H_2O_2 produced by SOD. This system produces a bactericidal agent in the vacuole.

MPO catalyses the oxidation of ubiquitous halide ions to produce hypohalite ions

$$X^- + H_2O_2 \rightarrow XO^- + H_2O$$

Cl^-, Br^- or I^- ions may participate in this reaction although Cl^- seems to be the physiological substrate, because it is found in abundance in the cells. The microorganism may be killed by halogenation of the cell wall or decarboxylation of its amino acids

$$R\text{—}CHNH_2\text{—}COOH \rightarrow R\text{—}CHO\text{—} + CO_2\,NH_3$$

The amino acid is converted to aldehydes, CO_2 and ammonia. Other, more potent ROIs may be formed in the phagocytic vacuoles, which also contribute to the bactericidal effect. It has been postulated that in a Fenton-type reaction

$$O_2^{-}\cdot + H_2O_2 \xrightarrow{\text{metal catalysis}} \cdot OH + OH^- \, O_2$$

and during the reaction with $O_2^{-}\cdot$ of alkyl hydroperoxides produced in the phagocyte membrane

$$O_2^{-}\cdot + R\text{—}OOH \rightarrow \cdot OH + OR^- + O_2$$

$\cdot OH$ is produced. In addition, $^1O_2$ may perhaps be formed instead of O_2 in the course of MPO activity, in the Fenton reaction, or during the spontaneous dismutation of oxygen (see Chap. I, Sect. 5). Part of the $O_2^{-}\cdot$ and H_2O_2 probably enters the cytoplasm from the phagocytic vacuole or may also get into the extracellular space. The cytoplasm is protected by its enzymes SOD, CAT and glutathione peroxidase (GP). The defence against the intermediates entering the extracellular space is less effective because of the lower activity of the extracellular scavengers (coeruloplasmin, transferrin, tetrameric SOD). Therefore, these intermediates play a role in the circumscribed inflammatory reactions occurring in connection with phagocytosis. The respiratory burst is a transient event lasting approximately 30-60 min. Its termination is partly due to the activity of the H_2O_2-MPO-halide system, which probably inactivates the NADPH oxidase system, and is partly the result of the internalization and degeneration of the NADPH oxidase system during phagocytosis [13, 15, 18, 51, 139, 147, 302, 467, 508, 525] (see Fig. 7).

Although they are not strictly physiological free radical reactions, we would like to discuss briefly some rare hereditary diseases the pathomechanism of which could be clarified with the help of what has been found out about the phenomenon of the respiratory burst. A disturbance in the O_2-dependent killing mechanism is a common feature of these diseases.

Chronic granulomatous disease (CGD) is one of this group. It is characterized by severe bacterial (mostly *Staphylococcus aureus*) and fungal infections occurring in childhood. The patients' leukocytes, although capable of normal phagocytosis, cannot kill the pathogens because of the absence of activation of the respiratory burst (Fig. 8).

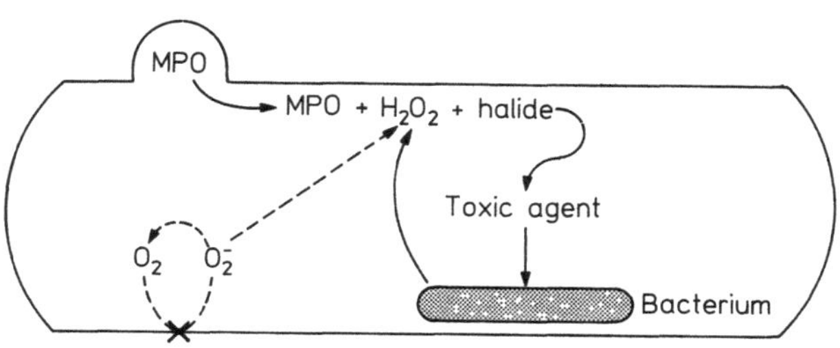

Fig. 8. Disturbed oxidative metabolism of phagocytes in chronic granulomatous disease MPO myeloperoxidase: *broken arrows* killing mechanism absent; *full arrows* killing mechanism present. (After Klebanoff [302])

Table 1. Molecular defects underlying chronic granulomatous disease

Molecular defect	Diagnostic test	Other dysfunctions	Susceptibility to infection	Inheritance
Defective initiation of oxidative metabolism				
1. Glyoprotein 150 deficiency	PAGE	Phagocytosis	Severe	AR
2. Trigger defect I	O_2^- · assay with	Degranulation	Severe	AR
3. Trigger defect II	OPZ and PMA	Chemotaxis	Severe	X-linked
4. Abnormality of membrane potential	Potential measurement with dyes		Severe	X-linked/AR
Defective NADPH supply				
5. G-6-PD deficiency	G-6-PD activity in red cells	Haemolysis	Normal to severe	X-linked
6. G-6-P translocase deficiency	Latent G-6-Pase activitiy in liver	Neutropenia Chemotaxis	Moderate to severe	AR
Defective activity of oxidative metabolism				
7. Low-affinity NADPH oxidase	O_2^- · assay with membrane fractions		Moderate	X-linked/AR
8. Cytochrome *b* deficiency				
- Without O_2^- production	Spectroscopy		Severe	X-linked
- with O_2^- production			Moderate	X-linked
Genetic linkage				
9. Kx deficiency	Reaction of red cells to anti-Kx	Acanthocytosis Haemolysis	Moderate to severe	X-linked

AR autosomal recessive; G-6-P glucose-6-phosphate; G-6-PD glucose-6-phosphate dehydrogenase; OPZ opsonized zymosan; PMA phorbol myristate acetate; PAGE polyacrylamide gel electrophoresis

At the same time, the leukocytes can kill some organisms, e.g. pneumococci and streptococci, because these produce H_2O_2, and the intact MPO system can destroy them. The disease is transmitted either as an X-linked trait or by autosomal recessive inheritance. Several kinds of molecular defects may be responsible for the absence of the activation of the respiratory burst and the consequent clinical symptoms. The NADPH oxidase system of phagocytes is

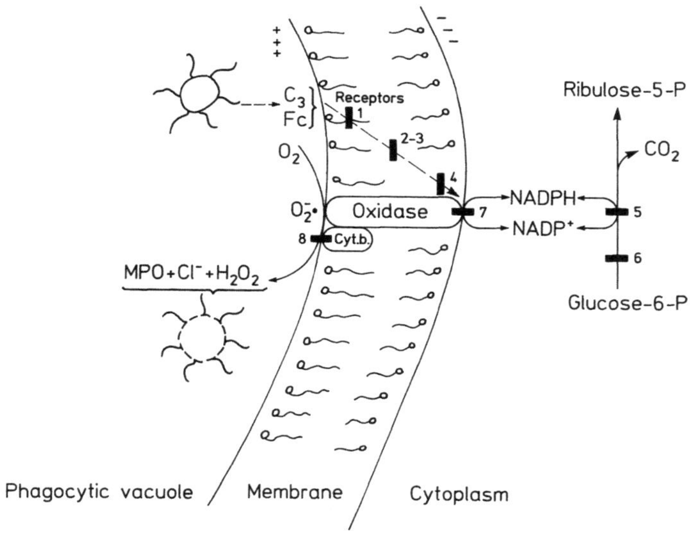

Fig. 9. Molecular defects causing chronic granulomatous disease. 1 glycoprotein-150 deficiency; 2-3 trigger defect I + II; 4 abnormality of membrane potential; 5 glucose-6-phosphate dehydrogenase deficiency; 7 low-affinity NADPH oxidase; 8 cytochrome *b* deficiency; MPO myeloperoxidase. (After Hitzig and Seger [259])

composed of three parts: an activating appratus, an electron transport chain containing flavoprotein and cytochrome *b*, and enzymes providing for the supply of NADPH (hexose monophosphate shunt). A defect of any of these components may cause the absence of activation of the system (Table 1, Fig. 9).

In a boy of CGD phenotype, the absence of a membrane glycoprotein was demonstrated, which was associated with severe damage of the C3- and Fc-receptor-mediated functions of phagocytes. In other words, this was a defect of the receptor binding function needed for membrane contact. Since the molecular weight of the glycoprotein was 150 daltons, it was called gp 150.

After the activating agent has established contact with the cell surface, a signal is propagated along the membrane. Two types of defects of signal propagation have been demonstrated: trigger defects I and II. As has been shown, the phagocyte membrane is depolarized in association with the activation of the cell. Abnormal changes in the membrane potential have been found in several patients with CGD of X-linked transmission, and in one case with CGD transmitted as an autosomal recessive trait.

Of the constituents of the NADPH synthesizing system, the deficiency of glucose-6-phosphate dehydrogenase (G-6-PD), known to cause haemolytic anaemia, may produce a clinical picture corresponding to CGD. Deficiency of another constituent of the system, glucose-6-phosphate translocase, may also produce the CGD phenotype. Low affinity of NADPH oxidase to NADPH may be another factor causing the defect.

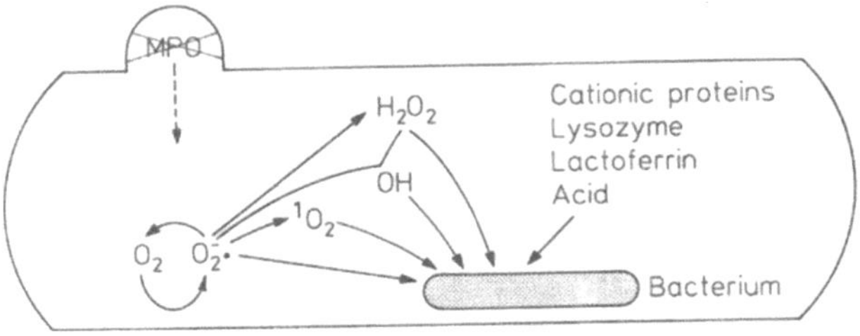

Fig. 10. Antimicrobial systems of myeloperoxidase-deficient neutrophils. (After Klebanoff [302])

Cytochrome *b* deficiency is the most frequent defect. Cytochrome *b* activity may be diminished or absent. In a few cases CGD was found to be associated with the very rare McLeod blood group phenotype, which is due to the absence of the X-linked Kx antigen (Kell system).

In hereditary MPO deficiency, too, leukocytes are unable to kill bacteria, but the consequences are less severe than in CGD. In this condition there is a period of latency in the bactericidal activity of leukocytes, followed by normal function, as opposed to CGD, in which leukocytes are completely deprived of their ability to kill bacteria. This is only a defect of early bactericidal activity, and the respiratory burst is normal. The bacteria are killed with the help of ROIs and oxygen-independent antimicrobial systems (acid, lysozyme, lactoferrin, cationic proteins), but only after some delay, indicating that MPO has a role in early bactericidal activity (Fig. 10). It follows that infections are, as a rule, less severe in patients with MPO deficiency than in CGD patients. Proneness to infections is enhanced by hypoxia and by an overactivation of the respiratory burst, since MPO also plays a role in the termination of the burst.

Patients with glutathione reductase deficiency have also been described. These were characterized by acute haemolysis due to the oxidant stress, by early cataract, and by the absence of an increased proneness to infections. Granulocyte function was normal, with the difference that the respiratory burst lasted only 5 min, during which time the phagocytosed bacteria were efficiently killed off [14, 259, 323, 487].

The absence of the activation of the respiratory burst may be responsible for the survival of intracellular pathogens in mononuclear phagocytes. For instance, *Toxoplasma gondii* may survive and proliferate in the macrophages. Only monocytes, the precursors of macrophages, are able to kill the organism, the macrophages themselves can only kill it in opsonized form. The survival of the organism in the macrophages is due to the absence of activation of the respiratory burst during its phagocytosis [584].

2 Free Radicals Generated During Platelet Activity

$O_2^-\cdot$ formation has been demonstrated in relation to thrombocyte activity. Its role, however, differs from that of the superoxide radical formed during the respiratory burst of phagocytes. ROIs produced in phagocytes serve to kill microorganisms, while $O_2^-\cdot$ in the platelets is probably a by-product of the arachidonic acid metabolism of these cells. Thus, although it has no specific function, when entering the extracellular space, it may damage neighbouring platelets, leukocytes and also the vascular wall. Stimulated platelets suddenly use oxygen for a very short time, which is probably spent entirely on the oxidation of endogenous free arachidonic acid, as opposed to phagocytes, which use most of the oxygen for ROI production. In platelets, $O_2^-\cdot$ is formed not in a burst, but continuously, and is independent of the cells' stimulation.

Free radicals and LPO products probably modulate platelet function by acting on the metabolism of the arachidonic acid of platelets. They accelerate the release of arachidonic acid from the phospholipids of the platelet membrane and shift the prostacyclin/thromboxane A_2 (PGI/TXA_2) balance in favour of TXA_2, enhancing platelet aggregation. Thus they have a thrombogenic function. Antioxidants (e.g. vitamin E) inhibit platelet aggregation by interfering with the mechanisms described, e.g. by shifting the PGI_2/TXA_2 balance in favour of PGI_2, which has an antiaggregation and vascular tone decreasing action (see Sect. 4.1) [225, 332, 340, 416].

3 Free Radical Generation in Mitochondrial and Microsomal Electron Transport Chains and in Peroxisomes

Mitochondria play an important role in the detoxication of oxygen by tetravalent reduction to water, mediated by the cytochrome oxidase system. This process is associated with the production of high-energy phosphate bonds, i.e. ATP. It has been shown that constant amounts of $O_2^-\cdot$ are continuously formed in the mitochondria. Constituents of the electron transport chain

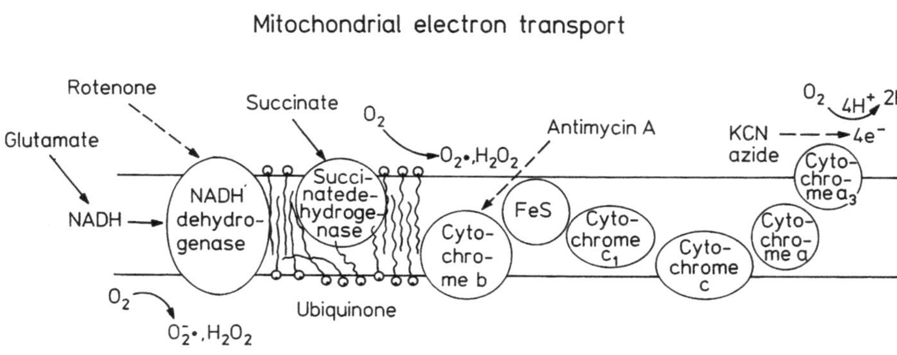

Fig. 11. Mitochondrial free radical production. The sources of mitochondrial $O_2^-\cdot$ production were investigated using various electron transport inhibitors (rotenone, antimycin A, KCN and azide: *broken lines*) and substrates (substrates bound to NADH and succinate: *full lines*). NADH dehydrogenase and dihydroorotic acid dehydrogenase and the ubiquinone-cytochrome *b* region can reduce oxygen to $O_2^-\cdot$, which is, at the same time, the precursor of H_2O_2. (After Freeman and Crapo [176])

(NAD, FAD, coenzyme Q, cytochrome *b*) localized in the inner membrane of mitochondria, if reduced, may react with O_2 to produce $O_2^-\cdot$. The ubiquinone (coenzyme Q)-cytochrome *b* region has been shown to be the main site of $O_2^-\cdot$ formation by autooxidation of ubisemiquinone. NADH dehydrogenase and dihydroorotic acid dehydrogenase are likewise autooxidable electron carriers responsible for part of the $O_2^-\cdot$ formation (Fig. 11) [176, 405].

In this way, there is a constant "univalent leakage" in the mitochondria using about 1%-2% of the electrons transferred to cytochrome oxidase for $O_2^-\cdot$ produktion. MnSOD found in the matrix can transform only about 80% of this univalent leakage to H_2O_2. Some 20% of the $O_2^-\cdot$ enters the cytoplasm, where it is neutralized by cytoplasmic SOD in association with the other defence systems.

These systems are able to control univalent leakage at normal oxygen tension, whereas hyperoxia increases the leakage. The leakage also increases in hypoxia, if there is not complete anoxia, because the members of the electron transport chain are reduced and become dissociated, and $O_2^-\cdot$ is formed as a result of their autooxidation. Thus, the 20% atmospheric oxygen seems to be an optimal value for univalent leakage. Finally, a role has been attributed to the $O_2^-\cdot$ generated in mitochondria in oxidative phosphorylation associated with the electron transport, i.e. in ATP synthesis, the exact mechanism of which is still unknown. ATP synthesis from ADP + $O_2^-\cdot$ via an ADP-ozonoid carrier intermediate has been postulated [176, 329, 405, 557] (Fig. 12).

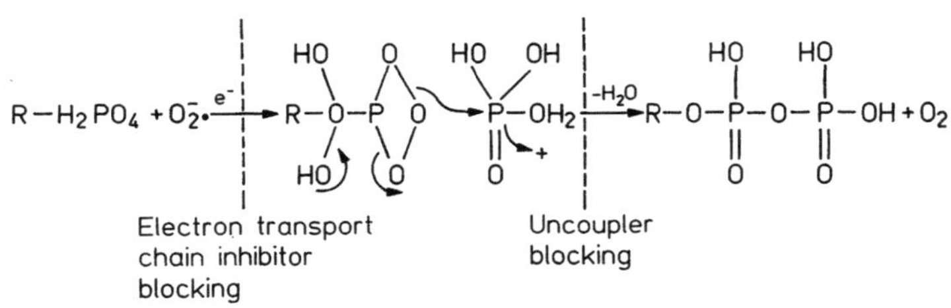

Fig. 12. The postulated role of $O_2^-\cdot$ in oxidative phosphorylation. (After Lippman [329])

ROIs may also be produced in the cytoplasm and nuclear membrane in electron transport chains unconnected with oxidative phosphorylation (cytochrome P-450, cytochrome b_5) Fig. 13)

Mixed function oxidases are responsible for cellular demethylation, hydroxylation and desaturation. In addition, they probably also play a role in the decomposition of the end products of LPO.

In the cytochrome P-450 system the decomposition of the P-450-Fe^{3+}-substrate-$O_2^-\cdot$ complex is the source of $O_2^-\cdot$, but it is mainly cytochrome P-450 reductase flavoprotein that produces $O_2^-\cdot$ by electron transfer from O_2 (Fig. 14).

We do not know why free radicals are formed in the interaction of some substrates with the cytochrome P-450 system, while others do not behave as

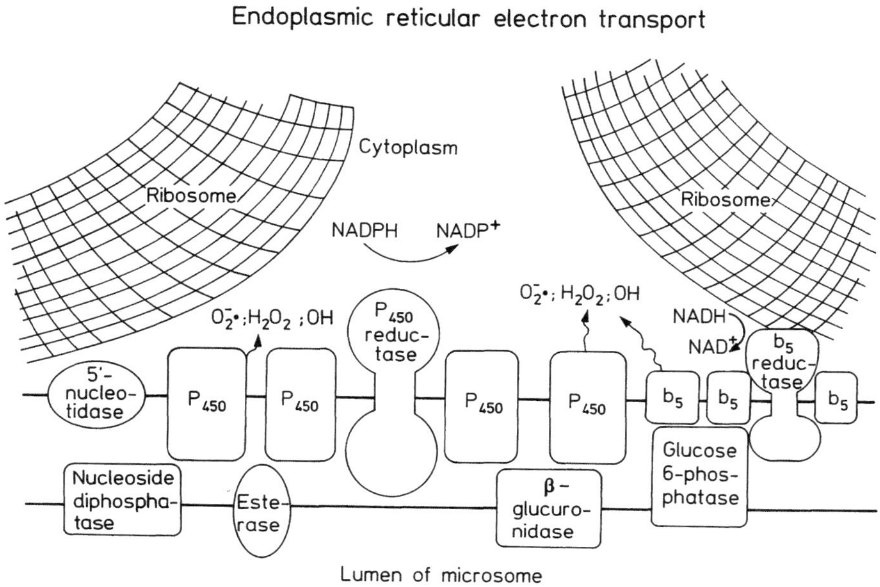

Fig. 13. Free radical formation in the endoplasmic reticulum. (After Freeman and Crapo [176])

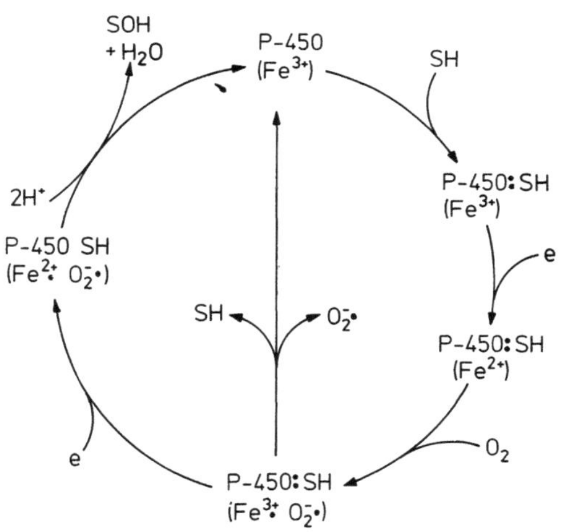

Fig. 14. Postulated mechanism of free radical formation in the cytochrome P-450 system. Ferricytochrome P-450 binds substrates or pseudosubstrates (SH) and a P-450-substrate (pseudosubstrate)- O_2 complex ist formed. If substrate is bound, the cycle is terminated by the formation of hydoxylated product (SOH) and water. In case of pseudosubstrate participation, however, the complex is decomposed and ferricytochrome P-450, $O_2^-\cdot$ and the pseudosubstrate are released. (After Hornsby [264])

inducers. The ROIs produced cause LPO which also damages cytochrome P-450. This mechanism may be responsible for the damage occurring during cytochrome P-450 substrate metabolism. This damage may be prevented by antioxidants, especially by those which can enter the active site of the enzyme (e.g. ascorbic acid, BHA). Figure 13 also shows that free radical mechanisms participate in the detoxification of xenobiotics by monooxygenases under normal conditions.

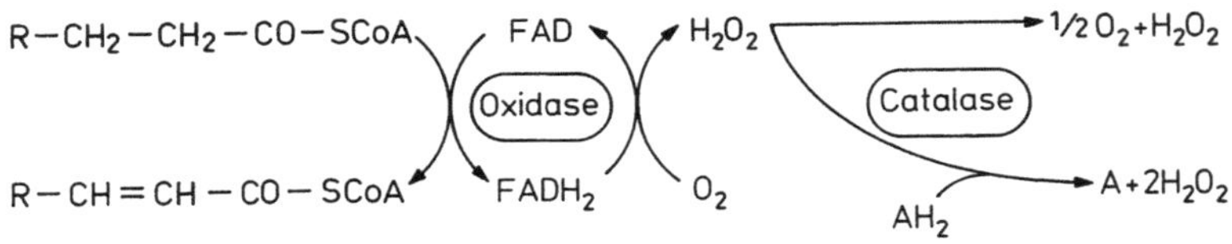

Fig. 15. Peroxisomal oxidation of fatty acids with the concomitant production of H_2O_2. A = oxidized substrate. (After Crane and Masters [91])

Owing to their high oxidase content (D-amino acid oxidase, α-hydroxy acid oxidase, fatty acid-coenzyme A oxidase), peroxisomes are the site of H_2O_2 production, which is then neutralized by CAT. It has been shown lately that the β-oxidation of fatty acids, which was earlier supposed to be an exclusively mitochondrial process in mammal cells, also takes place in the peroxisomes. Thus, peroxisomes play an important role in the balance of lipid metabolism. CAT can eliminate H_2O_2 produced during fatty acid-coenzyme A oxidase activity or in connection with the function of other oxidases by catalytic and peroxidase activity [91, 176, 263, 318] (Fig. 15).

4 Role of Free Radicals in the Synthesis and Degradation of Various Biomolecules: Free Radical Formation During Autooxidation of Biomolecules

4.1 Arachidonic Acid Metabolism

Arachidonic acid metabolism is, in fact, LPO, which, unlike the uncontrolled, nonenzymatic, pathological LPO, is a stereospecific, enzymatically controlled process. ROIs play a part in several steps of the arachidonic acid metabolism (Fig. 16). They increase the activity of membrane phospholipase A_2, releasing arachidonic acid from membrane phospholipids. Phospholipase A_2 is activated when LPO triggered by ROIs changes the phospholipid environment of the enzyme. Peroxidized phospholipids, being better substrates of the enzyme than their nonperoxidized counterparts, also contribute to the activation of phospholipase A_2.

Through their connection with arachidonic acid metabolism, free radicals may influence blood clotting (TXA_2/PGI_2, balance), inflammation (leukotrienes: 5-HETE, LTB_4, SRS-A), and the respiratory pathway (SRS-A) (bronchial asthma, chronic spastic bronchitis). Just as with hydroperoxide fragmentation in LPO, malondialdehyde (MDA) is also formed as a by-product of TXA_2 synthesis in prostaglandin synthesis. In this case, however, ROIs play a role not only in the initiation of arachidonic acid metabolism, but also in the synthesis of prostaglandin endoperoxide (PGG_2 and PGH_2) from arachidonic acid catalysed by PG endoperoxide synthetase (PGH synthetase), an enzyme with both cyclooxygenase (producing PGG_2 from arachidonic acid) and per-

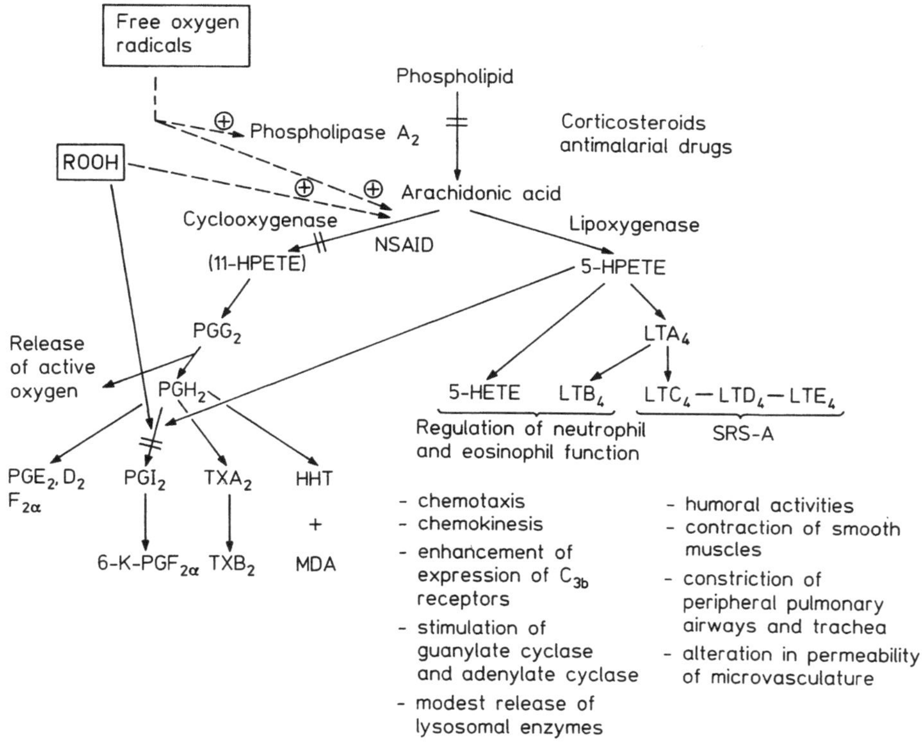

Fig. 16. Connection of ROIs with prostaglandin synthesis. NSAID non-steroid anti-inflammatory drug; TXA_2 thromboxane A_2; TXB_2 thromboxane B_2; HHT (12S)-12-hydroxy-5Z, 8E, 10E)-heptadecatrienic acid; MDA malondialdehyde; 6k-$PGF_{2\alpha}$ 6-keto-$PGF_{2\alpha}$; HPETE hydroperoxyeicosatetraenic acid; HETE hydroxyeicosatetraenic acid; LT leukotriene; SRS-A slow reacting substance of anaphylaxis. (Modified from Goetzl [203])

oxidase (producing PGH_2 from PGG_2) activity. For its activity PGH synthetase needs arachidonic acid, O_2, lipid hydroperoxide activator, and heme.

The enzyme has an interesting autocatalytic property: PGG_2, the product of its cyclooxygenase activity, is needed as a hydroperoxide for the maintenance of the enzyme's cyclooxygenase activity. Other hydroperoxides have a similar effect. At the same time, the enzyme's peroxidase activity serves to eliminate considerable amounts of lipid hydroperoxide, while other prostaglandins are formed. Whether lipid hydroperoxides stimulate one or the other activity of PGH synthetase depends on their concentration: at 10^{-9}–10^{-7} *M*, they stimulate cyclooxygenase activity, whereas concentrations of 10^{-7}–10^{-6} *M* favour the peroxidase activity of the enzyme. Organic free radicals R· are produced during the peroxidase activity, which give rise to ROI formation and pathological LPO (Fig. 17).

If the peroxidase activity dominates, cyclooxygenase activity is inhibited, partly because of the inhibitory effect of the relatively high hydroperoxide

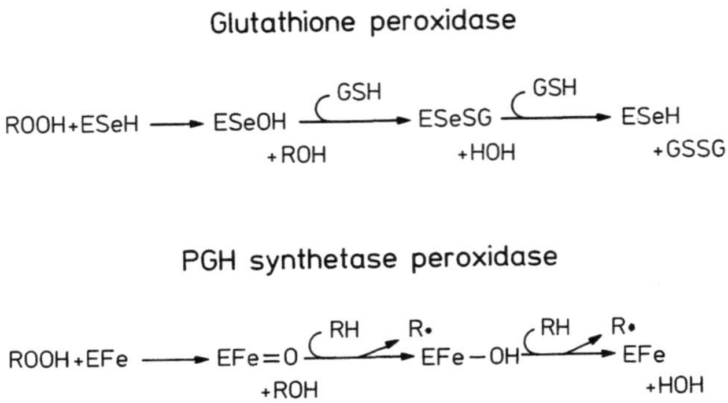

Fig. 17. Mechanism for removal of lipid hydroperoxides, for explanation see text. (After Lands) [314])

concentration needed for the stimulation of the peroxidase activity, and partly owing to the inactivating effect of the ROIs produced. The physiological concentration of hydroperoxides in the cell is probably 10^{-9} *M*, and the physiological stimuli of prostaglandin synthesis probably produce a concentration of 10^{-8}–10^{-7} *M*. Thus, under physiological conditions, the cyclooxygenase activity of the enzyme dominates, which eliminates peroxides, and neutralizes the PGG_2 formed during cyclooxygenase activity to such an extent that the intracellular hydroperoxide level is lower than necessary for the full activation of cyclooxygenase.

It is important that glutathione peroxidase decomposes hydroperoxides without producing free radicals, as opposed to the peroxidase activity of PGH synthetase. Thus, if hydroperoxide production in the cell attains a level (10^{-7} *M*) that can no longer be controlled by the defence mechanisms (glutathione peroxidase, glutathione-*S*-transferase, CAT), PGH synthetase starts functioning also as a peroxidase, producing ROIs and leading to pathological LPO. This is significant from the aspect of the mutagenic and carcinogenic actions, in which both the ROIs produced and the elevated cellular peroxide level play a role. These data furnish fresh evidence for the known association between chronic inflammation and carcinogenesis, and may, in part, explain the anticarcinogenic effect of some drugs inhibiting arachidonic acid metabolism (see also Chap. VI Sect. 10). As regards the other enzymes of arachidonic acid metabolism, PGI_2 synthetase is deactivated by micromolecular amounts of peroxides, whereas TXA_2 synthetase is completely resistant to deactivation by hydroperoxides. Therefore, if the intracellular lipid peroxide concentration rises above 10^{-7} *M* this not only results in the stimulation of peroxidase activity, but also produces a shift in the PGI_2/TXA_2 balance.

Lipid hydroperoxides are produced in the cell not only in LPO by virtue of the cyclooxygenase activity of PGH synthetase, but also mediated by the lipoxygenase enzyme. The activator lipid peroxides are needed for the function of this enzyme, too. Inhibition or congenital deficiency of lipoxygenase may lead to a decrease in activator peroxides to a level lower than necessary for the cyclooxygenase activity of PGH synthetases. Bleeding time is considerably prolonged in patients with congenital deficiency of platelet lipoxygenase, which is due to the following mechanism. The level of lipid hydroperoxide activators needed for PGH synthetase activity is low, and thus TXA_2, is not synthesized, which leads to deficient platelet aggregation.

Accordingly, the cyclooxygenase activity of PGH synthetase may be inhibited by: (a) antagonists competing for the substrate binding site of the enzyme; (b) noncompetitive antagonists attacking the binding site of the peroxide activator; (c) antioxidants interfering with the free radical chain reactions catalysed by the oxygenase; (d) irreversible damage of the active site of the enzyme; and (e) indirectly by lowering the level of the substrate arachidonic acid or of the lipid peroxide activator.

The nonsteroid anti-inflammatory agents are grouped according to the mechanisms described under (a), (b) and (d):

(a) competitive reversible antagonists
(b) noncompetitive reversible antagonists
(d) competitive irreversible antagonists

Phenol derivatives, e.g. acetamidophenol and phenylbutazone, are noncompetitive reversible antagonists. Their effect depends on the intracellular lipid peroxide level; they do not inhibit cyclooxygenase activity at high lipid peroxide levels, only at low levels. Therefore, although they have an analgesic effect at all lipid peroxide levels, their antiinflammatory effect is weak if the intracellular lipid peroxide level is high.

Vitamin E inhibits arachidonic acid release from the membranes and lipoxygenase activity, while it does not inhibit cyclooxygenase activity, and elevates the PGI_2 level because it protects PGI_2 synthetase against the inhibitory effect of lipid hydroperoxides (Fig. 18).

Depending on their concentration, antioxidants may inhibit or stimulate

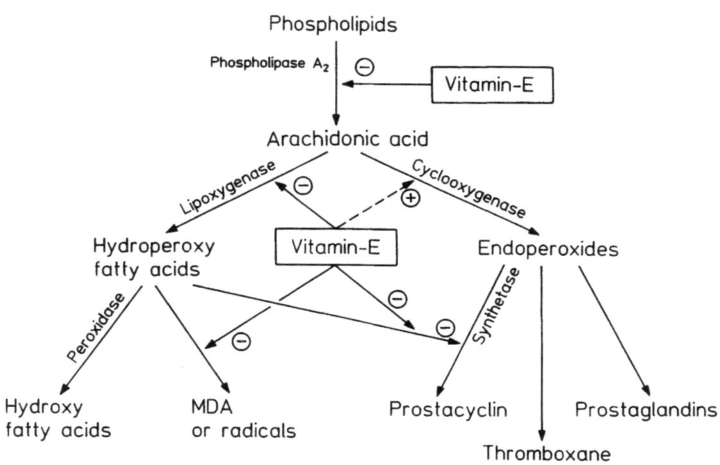

Fig. 18. Effects of vitamin E on arachidonic acid metabolism, for explanation see text. (After Panagamala and Cornwell [416])

cyclooxygenase activity, or they may leave it unaffected. High concentrations, as a rule, show an inhibitory effect, because they prevent the formation of intramolecular free radicals necessary for PG synthesis, and thus inhibit PG synthesis. Low antioxidant concentrations, however, protect cyclooxygenase against deactivation by preventing the excessive accumulation of (inhibitory) lipid hydroperoxides and by inhibiting the damaging effect of ROIs produced during the $PGG_2 \rightarrow PGH_2$ conversion [65, 108, 115, 313, 314, 416, 439, 488, 571].

4.2 Role of Free Radical Reactions in the Synthesis of Adrenocortical Hormones and in the Functional Division of the Cortex

The adrenal cortex is one of the organs best equipped with antioxidants (high SOD, glutathione, selenium, glutathione reductase, G-6-PD, vitamins E and C, and moderately high CAT and GP levels). Such a high degree of antioxidant protection indicates LPO in the organ even under physiological conditions, the antioxidants being necessary for the control and localization of LPO. Although we cannot embark here upon a detailed description of the biosynthesis of adrenocortical hormones, a few important steps will nevertheless be surveyed for a better understanding of the hypothesis referred to in the title of this section. The hydroxylations needed for the synthesis of adrenocortical hormones occur in the mitochondria and microsomes and are catalysed by the enzymes of the cytochrome P-450 system. P-450_{CMO} is a mitochondrial enzyme. It catalyses aldosterone synthesis from corticosterone by oxidizing the methyl group of C-18 to aldehyde (CMO = corticosterone methyl oxidase). Mitochondrial P-$450_{11\beta}$ (= 11-β-hydroxylase), microsomal P-450_{21} (= 21-hydroxylase) and 17-hydroxylase catalyse the synthesis of glucocorticoids (cortisol, corticosterone) (Fig. 19).

Aldosterone is produced exclusively in the zona glomerulosa, because the enzyme P-450_{CMO} is only found here. The zona glomerulosa does not contain 17-hydroxylase and, therefore, glucocorticoids and gonadotrophic hormones are not synthesized in this zone. At the same time, the zona fasciculata and zona reticularis contain 17-hydroxylase and, therefore, glucocorticoids and gonadal hormones are synthesized in them. It has been shown (see Fig. 13) that $O_2^-\cdot$ is released by the interaction of cytochrome P-450 with pseudosubstrates. According to the hypothesis of Hornsby and Crivello, the same occurs for the interaction of cytochromes participating in the synthesis of adrenocortical hormones with some steroids. The ROIs produced may damage cytochrome P-450 directly, or by inducing LPO. P-450 can use the lipid peroxides formed ROOH as oxygen donor substrates, while the alkoxy radical RO· is formed by the cleavage of the O-O bond. The reactivity of the alkoxy radical is almost as high as that of ·OH, and that is how cytochrome P-450 is deactivated. This mechanism may also play a role in the deactivation of P-450_{CMO}, P-$450_{11\beta}$ and P-450_{21}. It has been shown that in adrenocortical cell culture, various steroids reduce the activity of these enzymes; this effect can be inhibited by the addition of antioxidants. Of the three enzymes, P-450_{CMO} is the most

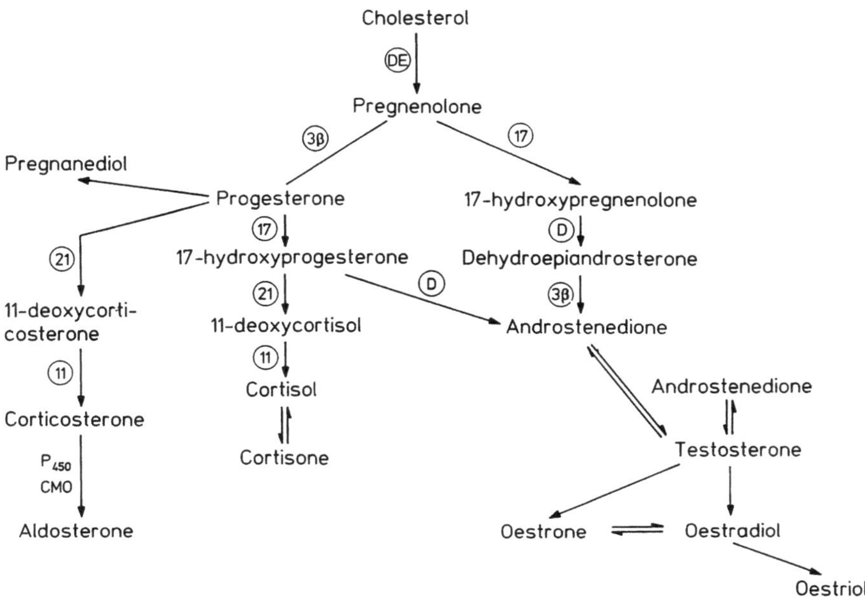

Fig. 19. Biosynthesis of steroids. DE debrancing enzyme; 3β 3-β-hydroxysteroid dehydrogenase; 17 17-hydroxylase; D $C_{17,20}$-desmolase (-liase); 21 21-hydroxylase; 11 11-β-hydroxylase; P_{450}-CMO corticosterone methyloxidase

sensitive to destruction by free radicals, followed by P-$450_{11\beta}$ and P-450_{21}. The enzyme 3-β-hydroxysteroid dehydrogenase may also be sensitive to LPO: its activity decreases in the adrenal cortex of animals with antioxidant (vitamins E and C) deficiency. As opposed to this, 17-hydroxylase is stable and remains ACTH-inducible even at high steroid levels. Probably LPO plays no important role in the absence of 17-hydroxylase from the zona glomerulosa. The so-

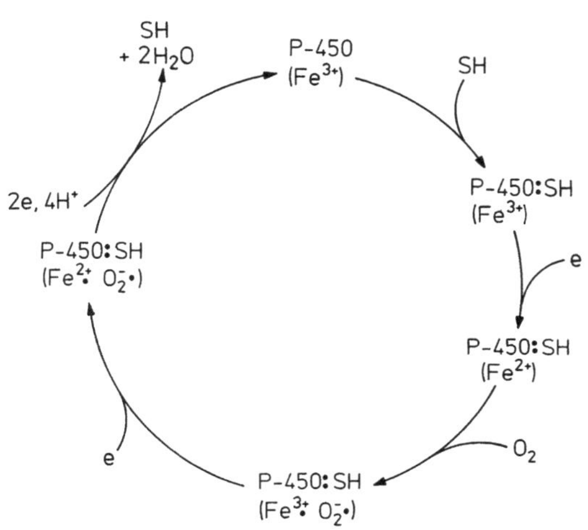

Fig. 20. Uncoupling of cytochrome P-450. If pseudosubstrates are bound, the substrate is not hydroxylated, the complex takes up two further electrons instead, and water and the original pseudosubstrate are released after fissure of the 0-0 bond. The two electrons are supplied by cytochrome b_5 via an NADH-dependent mechanism. (After Hornsby and Crivello [263, 264])

called uncoupling mechanism is more significant in the deactivation of P-450_{21} than LPO. In this mechanism the original substrate is released instead of being oxygenated, and water is produced (Fig. 20). Mostly C_{19} steroids (androgens), but also C_{21} steroids (corticosteroids) may act as uncoupling substrates.

The origin of the zones of the adrenal cortex has been studied extensively. In 1954 Bachmann [16] proposed that the cells of the outer cortex (the zona glomerulosa and the outer zona fasciculata) are the youngest, and it is in these zones that most cell divisions occur. The cells of the zona reticularis are the oldest, as shown by the accumulation of pigmented lipids in this zone. Cell death occurs predominantly in this region. According to this hypothesis, there is a constant flow of cells from the zona glomerulosa towards the zona reticularis, where the life cycle of the cells is terminated. This theory was later substantiated by the results of other authors. However, when it was discovered that the zones produce different steroids, it seemed difficult to reconcile this with Bachmann's hypothesis. It was, however, found that in culture, cells from rat and bovine zona glomerulosa took on characteristics of the zona fasciculata and zona reticularis cells, indicating that the two findings are not contradictory. It is, nevertheless, known that there is centripetal blood flow in the adrenals, which is directed from the capsule to the medulla. Thus, there is a steroid gradient with the steroid concentration increasing in the cortex from the outside towards the centre.

According to the hypothesis put forward by Hornsby and Crivello, the differences in steroid secretion between the zones of the adrenal cortex are due to variations in the distribution of the enzymes involved in steroid synthesis in the zones. During their life cycle the cells of the adrenal cortex migrating from the outer cortex in the direction of the zona reticularis meet with an increasing steroid gradient. In the course of the cytochrome P-450-steroid interaction, ROIs are produced, which cause LPO and eventually the various cytochromes are inactivated. P-450_{CMO} is the most sensitive to inactivation, therefore, it is found only in the outermost, thinnest zone, the zona glomerulosa. Since the structure of the steroids responsible for the inactivation of P-450_{CMO} differs only slightly from that of the steroids inactivating P-$450_{11\beta}$ and since both enzymes are found on the inner mitochondrial membrane, it is possible that the steroids interact directly only with P-$450_{11\beta}$, and the LPO thus produced inactivates P-450_{CMO}, too, which is found at the same site. Nevertheless, a direct interaction between the steroids and P-450_{CMO} cannot be excluded either. P-$450_{11\beta}$ is probably inactivated in vivo by corticosterone. Cortisol and androstenedione have similar effects, thus indirectly P-450_{CMO} is inhibited together with P-$450_{11\beta}$.

Since corticosterone is synthesized in the zona glomerulosa and also cortisol appears on the zona glomerulosa-zona fasciculata border, P-450_{CMO}, which is the most sensitive to peroxidation, has by now been completely inactivated and is absent from the zona fasciculata. This explains why aldosterone is only synthesized in the zona glomerulosa. LPO does not explain the absence of 17-hydroxylase activity from the zona glomerulosa, this has been attributed to phosphodiesterase activity being higher in the zona glomerulosa than in the zona fasciculata, because the induction of 17-hydroxylase by ACTH is mediated by cAMP, which is decomposed by phosphodiesterase. P-$450_{11\beta}$ is less sensitive to LPO than P-450_{CMO}, but its activity decreases gradually towards the zona reticularis. The same is true for P-450_{21} and 3-β-hydroxysteroid dehydrogenase. At the same time, 17-hydroxylase is also needed for the ACTH-induced synthesis of gonadal hormones. The activity of $C_{17,20}$ liase and sulphotransferase participating only in the synthesis of these hormones increases towards the inner zones of the adrenal cortex. Therefore, only a considerably reduced cortisol and corticosterone release is found in the zona reticularis as compared with the zona fasciculata, because cortisol itself and also androstenedione decrease P-$450_{11\beta}$ and P-450_{21} activity considerably. Androgenic steroids also inhibit P-450_{21} action by the uncoupling mechanism. Thus P-$450_{11\beta}$ and P-450_{21} sensitize the cells against the steroid gradient, and the cells produce different steroids in the course of their maturation (ageing).

LPO is responsible not only for the zonal distribution of hormone release, but also for the ageing of the cells migrating from the zona glomerulosa towards the zona reticularis, for the accumulation of pigmented lipids concomitant with the ageing process, and for the increased cell destruction in the zona reticularis. LPO is not prevented by the antioxidants which are in abundance in the adrenal cortex, it is only decreased to a very low level and is localized. This theory explains the function of the different zones of the adrenal cortex, which release specific hormones, and the presence of antioxidants. It does not, however, give a reason for the adrenocortical 17-hydroxylase levels and the distinction it makes between the zona fasciculata and the zona reticularis is not as clear-cut as that made between the zona glomerulosa and the zona fasciculata.

It is interesting to note that the ACTH-induced ascorbic acid depletion used for ACTH bioassay is probably also due to free radical reaction. ACTH increases cytochrome P-450 activity, and the consequent release of ROIs and lipid radicals leads to the conversion of ascorbinic acid to dehydroascorbinic acid and monodehydroascorbinic acid. These, unlike ascorbic acid, do not remain in the cell.

o,p'DDD,(2,2) 4-chlorophenyl, 2-chlorophenyl (1-dichloroethane), used for the treatment of metastases of adrenocortical carcinoma, inhibits the activity of mitochondrial P-450 (P-$450_{11\beta}$, P-450_{CMO}) and thereby steroid release. There are data indicating that this effect is due to its LPO-inducing effect, similarly to the insecticide DDT, its chemical predecessor [94, 264]. Doxorubicin has also been shown to have a ROI-producing effect (see Chap. U Sect. 12), whereby it inhibits P-$450_{11\beta}$ activity [99].

4.3 The Role of Free Radical Reactions, in the Function of Melanins

Melanins are naturally occurring redox polymers derived from tyrosine, which contain high concentrations of *ortho*-hydroquinone and *ortho*-quinone units. Some of them interact with each other to form quinhydrones (~Q...HQ~).

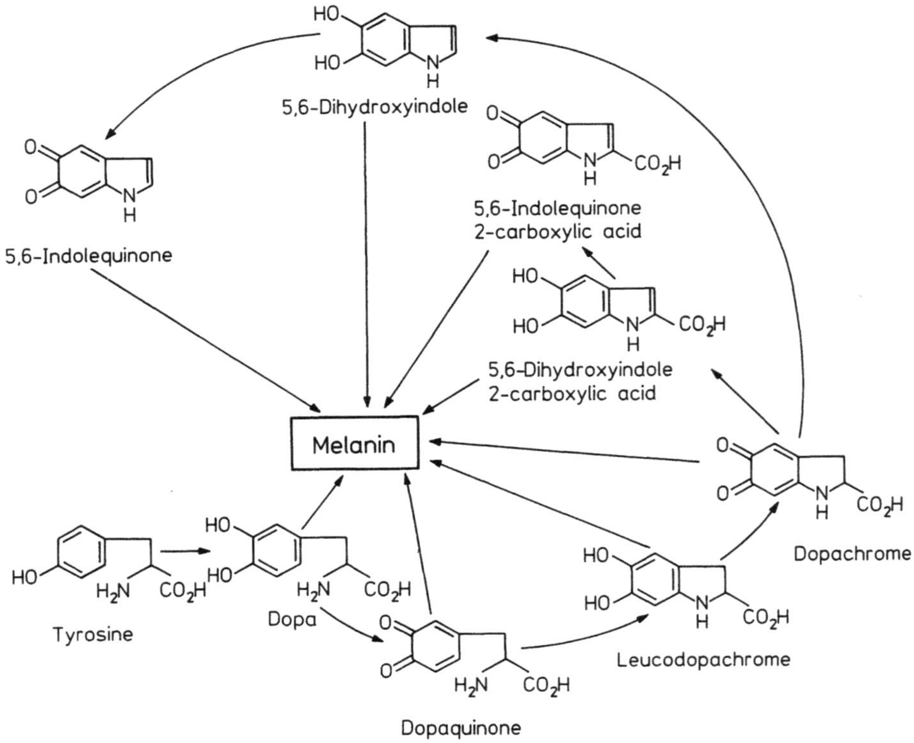

Fig. 21. Production of melanin from tyrosine. (After Sealy et al. [486])

The initial oxidative steps of the synthesis to dopaquinone are mediated by the copper-containing enzyme tyrosinase (Fig. 21). Dopa probably accelerates the enzymatic oxidation of tyrosine. An alternative pathway has been proposed in which oxidation of tyrosine is catalyzed by H_2O_2-peroxidase. The subsequent steps occur spontaneously and do not require enzyme mediation. Melanin is a heterogeneous polymer in which all the intermediates of melanogenesis between dopa and indolequinone are incorporated. It contains several units in quinone (⁓Q) and hydroquinone (⁓HQ) oxidation states and that is why it is a redox polymer.

The presence of intrinsic free radicals has been shown in melanin, which probably correspond to semiquinone ($\sim Q^{-\cdot}$) radicals. Some of these radicals are inert chemically, while others are in equilibrium with the quinone and hydroquinone units, i.e. with the oxidized and reduced forms of semiquinone. Melanins contain associated metal ions (Cu, Zn, Fe), water and protein (part of this may be the enzyme tyrosinase).

In addition to the intrinsic free radicals, melanin also contains transient free radicals, since it absorbs light in both the UV and visible spectra. Free radical mechanisms may account for part of the photoprotective process believed to be the primary function of melanin. Upon light absorption melanin

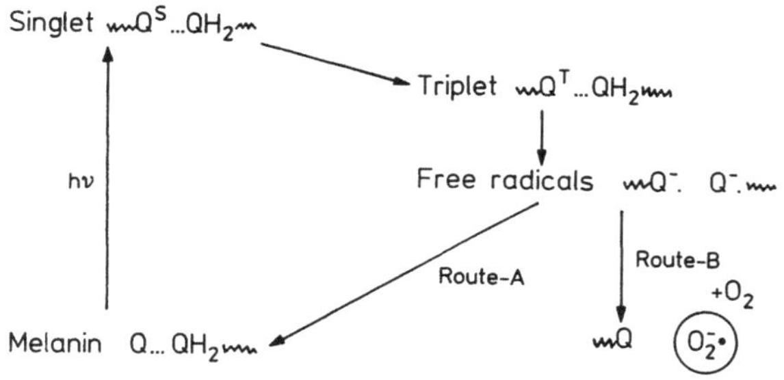

Fig. 22. Production of free radicals on light absorption by melanin, for explanation see text. (After Sealy et al. [486])

is raised to an excited state. The excitation energy is localized in the quinone moiety (Fig. 22). As a result of electron transfer among semiquinones, the system returns to its original state. A minor pathway (route B) is also possible in the presence of O_2. Here chemical reaction occurs and some hydroquinone groups are oxidized to quinones while $O_2^-\cdot$ is produced. Thus potentially damaging ROIs are generated upon the photoirradiation of melanin.

Route B leads to partial oxidation of melanin owing to the $\sim QH_2 \rightarrow \sim Q$ conversion [486]. Melanins may produce ROIs not only by light absorption, but also in the dark, in the presence of oxygen; the production of $O_2^-\cdot$ and OH has been observed [93]. Melanin has been found to neutralize exogenous free radicals formed either by photolysis or radiolysis. It has been suggested that melanin may be a good ·OH, H·, e_{aq}^-, alkyl, etc. scavenger. Strong binding of some drugs, e.g. cocaine, adrenaline, dopamine, and phenothiazines to melanin has been demonstrated. Most of these drugs are supposed to form charge-transfer complexes with melanin, in which the drug acts as an electron donor and melanin as an electron acceptor. This may explain observations such as the frequent occurrence of cutaneous photosensitivity in patients taking phenothiazines [195, 486].

4.4 The Modulating Effect of Free Radicals on the Activity of Enzymes: Free Radical Generation by Autooxidation of Biomolecules and During Enzyme Function

$O_2^-\cdot$ for instance, inactivates several enzymes, e.g. papain, yeast alcohol dehydrogenase, glycerinealdehyde-3-phosphate dehydrogenase, ribonuclease, yeast lysine and RNA-ligase. At the same time, some enzymes require $O_2^-\cdot$ to be activated, e.g. 2-nitropropane dioxygenase, indolamine-2-,3-dioxygenase, dopamine-β-dihydroxylase, and indophenol dioxygenase. The two last-named enzymes may be involved in the pathogenesis of schizophrenia, paranoia, hallucination and Korsakoff's syndrome, indicating the possible involvement of free radical reactions (see Chap. VI Sect. 5) [404, 500].

Autooxidations in biological systems proceed via the easier route of univalent reduction, as a result of which $O_2^-\cdot$ is formed. The oxidized forms of haemoglobin and myoglobin are, in fact, not in the ferrooxy state as was believed earlier, but in the ferrisuperoxy state, and thus the possibility of $O_2^-\cdot$ release exists. $O_2^-\cdot$ is also produced in the autooxidation of oxyhaemoglobin to methaemoglobin. In addition, $O_2^-\cdot$ generation is unavoidable in the course of the function of various enzymes, e.g. xanthine oxidase, aldehyde oxidase and dihydroororotic acid dehydrogenase, which generate significant amounts of $O_2^-\cdot$, while flavoenzyme dehydrogenases produce smaller amounts [177, 179, 360]. A role may be attributed to the autooxidation of catecholamine in catecholamine cardiomyopathy; to $O_2^-\cdot$ generated from haemoglobin in some types of hemolytic anaemia; to xanthine oxidase in ischaemic changes; and to 6-hydroxy-dopamine (6-OHDA, for further explanation see note 5) autooxidation in some diseases of the central nervous system.

5 The Role of Free Radicals in Reproduction and Embryonic Development

It has been shown that a process similar to phagocytosis occurs during the fertilization of the eggs of sea urchins; peroxidation is triggered by specific stimuli. O_2 consumption increases in the fertilized egg, and an $O_2 \rightarrow H_2O_2$ conversion occurs. The peroxidase (ovoperoxidase) contained by the surface granules of the egg leaves the egg by exocytosis and with the H_2O_2 produced and with halide forms a system similar to MPO. Iodine, present in considerable amounts in sea water, is the halide substrate. The enzyme iodinates the cell membrane tyrosines and mediates di- and trityrosine crosslinking, leading to the hardening of the fertilization membrane. Thereby polyspermia is prevented. The continuous activity of the system probably also has a spermicidal function [303].

SOD activity has been demonstrated in human semen. It is in direct relation with the motility of spermatozoa. Rabbit spermatozoa have been found to undergo spontaneous LPO, which has been attributed to the $O_2^-\cdot$ they produce. Thus SOD provides protection for spermatozoa [261, 402].

The environment of the embryo in the uterus is probably hypoxic, because the defence systems against high O_2 concentrations have not developed at this stage. This concept is supported by the increasing levels of antioxidant enzymes observed late in gestation. The physiological stimulus of this increase is not known; it can only be postulated that the increasing metabolic activity of the lungs, or the increasing total O_2 consumption of the growing fetus might have such a role [175]. A progressive increase in placental SOD, CAT and peroxidase activity has also ben observed during gestation; this might be an important factor in the protective role of the placenta [351, 551]. Placental LPO may provide protection against potentially damaging free radicals. Unlike the activity of protective enzymes, LPO in the placenta is most active in the early phases of gestation, decreasing to a very low level by the end of pregnancy. Probably in the early phases increased placental LPO provides the protection, which is then taken on by the protective enzymes [551]. During normal pregnancy LPO may be observed in the plasma, too; its increase is directly related to the time of gestation and it shows a fall after birth. The activating effect of LPO on arachidonic acid metabolism may play a role in the induction of labour, while its PGI_2 synthesis inhibiting effect may lie behind thromboembolic complications and hypertension during pregnancy [607].

6 Summary

Physiological free radical reactions occurring in all parts of the cell play a role in a number of physiological processes. ROIs play an important part in the oxygen-dependent killing mechanism of phagocytes. In response to the appropriate stimulus, a coordinated metabolic chain reaction occurs in the phagocyte, which is called a respiratory burst. The purpose of this burst is to supply ROIs for the killing of the target cells of phagocytes. The description of oxygen-dependent killing has led to the understanding of the pathomechanism of some rare hereditary diseases characterized by disturbed leukocyte function (e.g. chronic granulomatous disease).

ROIs are also generated in association with platelet function, but these are probably only by-products of the arachidonic acid metabolism of platelets and not directly involved in platelet function. They are continuously produced and not dependent on stimulation. Free radicals shift the prostacyclin/thromboxane A_2 balance in favour of thromboxane A_2, thus enhancing platelet aggregation and thrombus formation.

In the mitochondrial electron transport chain ROIs are inevitably produced during oxidative phosphorylation: approximately 1 %–2 % of the electrons transferred to the enzyme cytochrome oxidase are involved in ROI formation. This "univalent leakage" increases in response to both hyperoxia and hypoxia. Reactive intermediates are also produced in electron transport chains in the cytoplasm and nuclear membrane, independent of oxidative phosphorylation (cytochrome P-450, cytochrome b_5). H_2O_2 is formed during peroxisomal activity; it is neutralized by CAT found in the peroxisomes.

ROIs and LPO caused by them interfere with arachidonic acid metabolism at several points, and ROIs are also released during the metabolism of arachidonic acid. The phopholipase A_2 activating effect of free radicals and the inhibitory effect of lipid hydroperoxides on prostacyclin synthesis are of major importance. ROIs may be released by the activation of PGH synthetase during arachidonic acid metabolism.

The adrenal cortex is well supplied with antioxidants, therefore, LPO is thought to play an important part in its physiological processes. Although the hypothesis of Hornsby and Crivello has not been proved, it provides a convincing explanation of the development of the zones of the adrenal cortex and the presence of high antioxidant concentrations.

Melanins are redox polymers with the primary function of photoprotection. Free radical mechanisms (scavenger effect of the melanin free radical) are supposed to play a role in this function. Potentially damaging ROIs are, however, produced during light absorbtion by melanin, and, in the presence of oxygen, also in the dark.

Free radicals influence the activity of enzymes: they inactivate some enzymes, while others need ROIs for their activation. ROIs are generated during the autooxidation of biomolecules and during enzyme activity.

There are data available indicating that free radical reactions play a role in reproduction and embryogenesis. LPO and the defence systems are involved in the function of spermatozoa, and are supposed to play a role in fetal development and the induction of labour.

Chapter IV

Pathological Free Radical Reactions: General Aspects

Uncontrolled, abnormal radical reactions occurring in the cell are called pathological free radical reactions. Such reactions may also be elicited by radicals generated under physiological conditions if the control mechanisms fail for some reason. They may further be caused by exogenous agents and physical interactions (ionizing radiation, UV, plastic implants).

1 Damage of Enzymes and Proteins Caused by Free Radicals

Polymerization, breakage of polypeptide chains and changes in the amino acid structure may occur in proteins and enzymes exposed to LPO in aqueous solution. Proteins containing unsaturated and sulphur-containing amino acids (tryptophan, tyrosine, phenylalanine, histidine, methionine, cysteine) are mostly sensitive to modification of the amino acid structure [176]. Proteins are also membrane constituents, therefore their damage explains the membrane-damaging effect of free radicals, while the loss of the specific activity of enzymes may also have severe consequences [535].

Some of the important membrane proteins are phospholipid dependent, their active configuration is generated via interaction of the hydrophobic parts of the polypeptide chain and the fatty acid parts of the special membrane phospholipids. Na^+, K^+-ATPase, succinyl dehydrogenase and cytochrome oxidase belong among these enzymes. Understandably, LPO triggered by free radical reactions inactivates these enzymes, as has been shown experimentally [115].

2 Damage of the Cell Membrane and Subcellular Organelles

Biomembranes and subcellular organelles are the chief targets of LPO-induced damage. Mitochondrial and microsomal membranes contain relatively more PUFA in their phospholipids than the plasma membrane, and are, therefore, more susceptible to peroxidation. Substances generating radicals damage various subcellular structures selectively, e.g. the damaging effect of CCl_4 is due mainly to the peroxidation of microsomes. ROIs produced in the microsomal electron transport chain during drug metabolization cause selective damage.

Damage of the lysosome membranes may lead to intracellular catabolism.

Lysosomes contain relatively less PUFA compared with the mitochondria and microsomes, and, therefore, they are less susceptible to injury [535].

3 Fluorescent Molecular Damage and Lipofuscin Pigments

Lipofuscin pigments accumulate in human and animal tissues as a result of ageing. Recent studies have shown that they are lipid-protein complexes resulting from the peroxidation of the PUFAs of subcellular membranes and correspond to the so-called residual body, the end product of lysosomal digestion. The pigments have characteristic fluorescent spectra. The Schiff base product of MDA (one of the terminal products of LPO), cross-linking with the primary amino groups of proteins and with the amino groups of nucleic acids and nucleic acid bases and phospholipids

$$RN{=}CH{-}CH{=}CH{-}NH{-}R$$

has similar fluorescent characteristics. This conjugated imine structure, having fluorescent chromophore properties, is an indicator of the cross-linking reactions of LPO. Thus the fluorescence method is suitable for the in vivo study of LPO damage [176, 261, 535].

MDA may be present in lipofuscin either in the Schiff base form, or in polymerized form. Ceroid is another lipopigment having a structure similar to that of lipofuscin, but unlike lipofuscin, it does not accumulate during normal ageing, only as a result of pathological processes [564].

4 Free Radical Damage of Nucleic acids and Their Components

Free radical damage of the DNA of living organisms may be due to breakage of the main chains (or of a strand), degradation of bases, or cleavage of hydrogen bonds. All components of nucleic acids may be exposed to free radical damage, which may become permanent, or may be repaired by special mechanisms. Unrepaired damage in bases may lead to mutation, while damage in the pentose part, to chain breakage. It is not known which of the DNA bases is the most sensitive to free radical reactions. Both thymine - from which the 5,6-dihydrothymine-5-il radical is formed - and guanine have been suggested.

Histones with high arginine and lysine content have been found to be good ·OH scavengers. Their role has not been clarified, but they may be responsible for the protection of DNA against pathological free radical reactions [49, 176, 391, 395, 600].

5 Connection Between LPO and Tissue Damage

LPO may proceed in a controlled form without causing cell damage, but it may also lead to tissue damage - either reversible or irreversible. In the latter case it causes the death of the cells (Fig. 23).

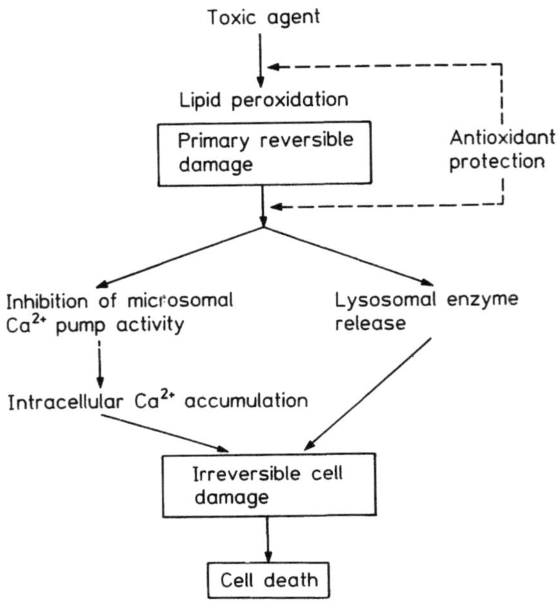

Fig. 23. Proposed sequence of events leading to cellular death after lipid peroxidation. (After Younes and Siegers [598])

According to Younes and Siegers, reversibility is due to cellular antioxidant protection. Lysosomal enzyme release invariably causes cell damage, but cell death is due to intracellular Ca^{2+} accumulation caused by the failure of the microsomal Ca^{2+} pump. It is important that LPO may not only be the cause of cell damage, but may also be the result of tissue damage caused by some other factor. It is thus hard to judge in some disease states whether cell damage due to LPO is primary, or whether LPO is secondary to cellular damage. Antioxidants are inactivated in the damaged tissue, and metals stored in the cells (mostly iron and copper) are released (e.g. due to decomposition of metalloenzymes by lysosomal proteases). These may lead to LPO [221, 598].

6 Summary

Pathological free radical reactions are uncontrolled abnormal free radical reactions occuring in the cell. They may be elicited by physiological free radicals if the control mechanisms are defective for some reason, or if there is a strong increase in the pro-

duction of free radicals under normal conditions of self-protection. Free radical reactions may damage any biomolecule: enzymes, proteins, carbohydrates, or nucleic acids. The genetic material, membranes composed predomintantly of lipids and proteins, and subcellular structures (plasma membrane, mitochondria, microsomes and lysosomes) are the main sites of LPO damage. Lipofuscin, a lipopigment, accumulates in the ageing human and animal organism; it corresponds to the residual body, the end product of lysosomal digestion. Lipofuscin is a product of LPO and thus its presence is evidence for LPO occurring in vivo. Tissue damage caused by LPO may be reversible if there is sufficient antioxidant protection. If this is depleted, the damage becomes irreversible. Cell death is probably due to accumulation of Ca^{2+} *caused by the failure of the* Ca^{2+} *pump. Secondary LPO occurs owing to membrane damage. This is hard to distinguish from primary LPO caused by free radical reactions.*

Chapter V

Demonstration of Free Radical Reactions

In this chapter only a very brief survey of the methods suitable for the demonstration of free radical reactions will be given, which is justified on the one hand by our scanty knowledge of the physicochemical background, and on the other, by the practical purposes of our book.

ROIs may be demonstrated directly by electron spin resonance (ESR) spectroscopy, by the spin trap method, with the help of more or less specific scavengers, and by chemiluminescence. Free radicals have a weak magnetic field on account of their unpaired electrons, which can be demonstrated by ESR spectroscopy. This is a very sensitive method suitable for demonstrating free radicals present in 10^{-5}–10^{-7} *M* concentration. If the free radicals are produced at a slow rate, or if their half-life is short owing to their high reactivity, they cannot be detected by ESR spectroscopy. In such cases the spin trapping method is used, which is, in fact, indirect ESR spectroscopy. In spin trapping the spectrum of a known stable compound is changed by the trapping of an unstable radical. The structure of the radical trapped is deduced from the ESR spectrum of the resultant relatively stable product [142, 280, 281, 344, 346].

Cholesterol, β-carotene and vitamin E are specific scavengers of singlet oxygen. $O_2^-\cdot$ can be detected by its reducing effect on ferricytochrome *c* and nitro blue tetrazolium, if it behaves as a reducing agent, and by the oxidation of adrenaline if it behaves as an oxidant. If a reaction can be inhibited with SOD, this may be a sign of $O_2^-\cdot$ generation. Ethanol, mannitol, *t*-butanol and isopropanol are scavengers of ·OH [172, 351, 500]. If a "specific" scavenger inhibits a reaction, it can be supposed that the free radical it scavenges participates in the reaction. However, these scavengers are insufficiently specific, and the inhibition they cause is not sufficient proof for the presence of the free radical. At the same time, the absence of the protective effect of the scavenger does not prove that the free radical is not involved in the reaction, because the absence of the scavenger-radical interaction may have steric reasons, namely that the scavenger cannot reach the site of free radical production [176].

Dark chemiluminescence (not visible to the naked eye) is probably a sign of oxidative reactions in living systems. Chemiluminescence is produced by singlet oxygen which either transfers its surplus energy directly by light emission, or reacts with a double bond, producing an excited carbonyl group RO*, which is then decomposed, accompanied by light emission. Thus chemiluminescence is suitable for demonstrating ROI production directly, and also for the detection of LPO [46, 58, 328]. The interactions of free radicals and

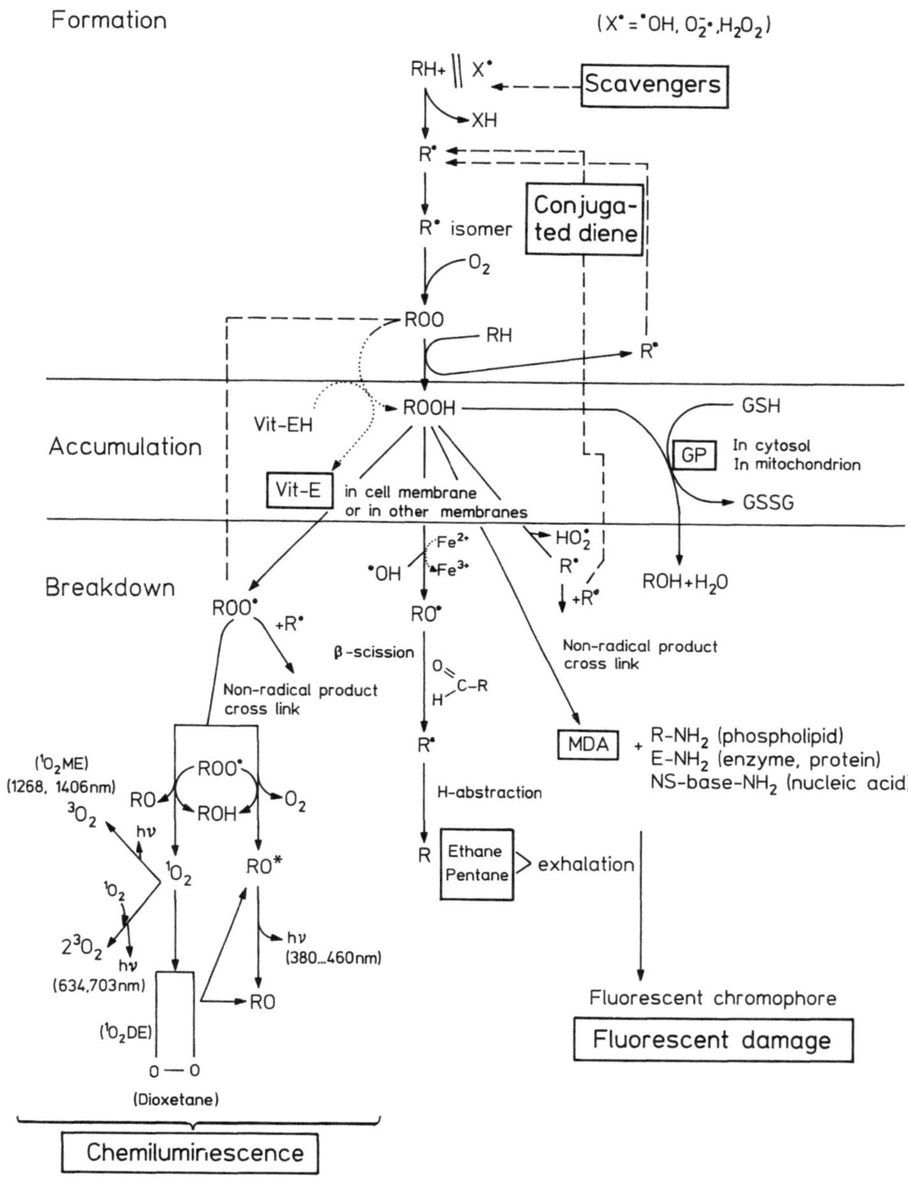

Fig. 24. Possibilities for the demonstration of free radical reactions and the consequent lipid peroxidation. ROOH lipid hydroperoxide; RO alkoxy radical; Vit-EH reduced vitamin E; Vit-E oxidized vitamin E; $^1O_2$ME singlet oxygen monomol emission; $^1O_2$DE singlet oxygen dimol emission; RO* excited carbonyl group, for further explanation see text. (After Boveris etal. [46])

also their physicochemical properties can be studied by pulse radiolysis, i.e. by short-term exposure to high-energy irradiation. If the chemical system irradiated is properly chosen, specific ROIs can also be produced in addition to the primary radical products generated by irradiation [12]. High-pressure liquid chromatography (HPLC) is a sensitive method, suitable for the electrochemical detection of ROIs [169].

LPO, the result of free radical reactions, is detectable by several methods. The UV spectrophotometry of conjugated dienes, the iodometric method, and the reduction of hydroperoxide catalysed by peroxidase are used for detecting lipid hydroperoxides produced during LPO [444]. The thiobarbituric acid (TBA) reaction is widely used for the detection of MDA, although the reliability of this method has recently been queried. It is based on the acid catalysed decomposition of lipid peroxide, in the course of which MDA is produced. The reaction of MDA with TBA produces red chromogen. However, MDA is not formed in the acid reaction of all lipid peroxides, which then remain undetected, while other biological substances give pseudopositive reactions. Therefore, HPLC is more suitable for the detection of MDA in biological samples [67, 144].

Fluorescent molecular damage can be studied by spectrophotometry [176, 352, 435, 531]. Ethane and pentane exhaled in breath give information on lipid peroxidation. Measurement of ethane seems to be more reliable, since, unlike pentane, it is not metabolized in the liver [315, 389, 536]. Decrease of the membrane PUFA content and of antioxidant (e.g. vitamin E, reduced glutathione) levels are indirect measures of LPO. Methods suitable for detecting LPO in vivo include ESR spectroscopy, demonstration of conjugated dienes, measurement of ethane and pentane in breath, demonstration of fluorescent damage and chemiluminescence. The levels of protective enzymes (SOD, GP, CAT) have also been used in experimental studies as indirect indices of free radical reactions, or for assaying antioxidant effect. Enzyme activity can be tested by chemical means when the catalysed reactions are observed, or by immunological techniques. Discrepancy between the results of the biochemical tests used for detecting enzyme activity and those of immunological tests measuring enzyme levels indicates the presence of inactive enzyme molecules [46, 142, 167, 168, 176, 280, 315, 328, 351, 435, 500, 531, 536, 538] (Fig. 24).

1 Summary

For direct detection of ROIs electron spin resonance (ESR) spectroscopy, spin trapping, high-pressure liquid chromatography (HPLC) and the more or less specific scavengers are used. Chemiluminescence is used for the direct and indirect detection of singlet oxygen. Free radicals are detected indirectly by demonstrating LPO. Lipid hydroperoxides produced during LPO are detected by UV spectrophotometry of conjugated dienes, iodometry, and by the peroxidase-catalysed reduction of hydroperoxides. Demonstration of MDA using the tiobarbiturate acid reaction is insufficiently reliable, therefore the use of HPLC is preferred. Spectrophotometric analysis of fluorescent molecular damage and the measurement of ethane and pentane in breath may give information on the intensity of LPO. Indirect conclusions regarding LPO can be drawn from the decrease of the cell membrane PUFA content, from the decrease of antioxidant level at the reaction site and from changes in the activity and/or level of protective enzymes.

Chapter VI

Clinical Importance of Free Radical Reactions and Their Role in the Pathogenesis of Various Human Diseases

1 The Radical Theory of Oxygen Toxicity

In normobaric hyperoxia, the most severe lesions are found in the lungs, whereas in hyperbaric hyperoxia, the damage of the central nervous system is even more severe and convulsions and grand mal epilepsy lead to death sooner than the pulmonary lesions would. In the pulmonary process, an acute or exudative phase and a subacute or chronic proliferative phase can be distinguished. The acute phase is characterized by perivascular, peribronchiolar, interstitial and alveolar oedema and alveolar haemorrhage with more or less extensive necrosis and the occurrence of type I epithelial cells. The proliferative phase is characterized by the progressive resorption of the exudates and thickening of the alveolar septa owing to hyperplasia of the interstitial cells. The type I alveolar cells are replaced by the relatively O_2-resistant type II cells. Collagen and elastin deposition in the interstitium contributes to the thickening of pulmonary tissue involved in gas exchange, causing disturbances of diffusion. Oxygen toxicity on the molecular level is attributed to the reaction of oxygen free radicals with cellular components. The generation of these free radicals is accelerated in hyperoxia, and damage occurs as soon as the defence mechanisms become insufficient. In the lungs, the endothelium

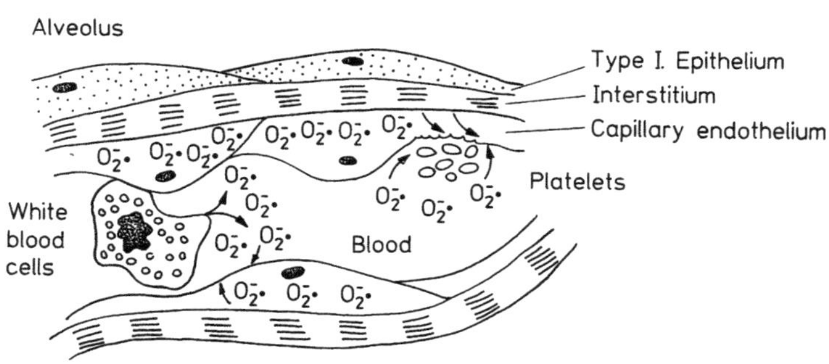

Fig. 25. Two-step mechanism of pulmonary endothelial damage caused by oxygen toxicity. 1. Intracellular $O_2^-\cdot$ production increases owing to hypoxia. 2. A chemotactic factor is formed from a plasma precursor under the influence of $O_2^-\cdot$ released from phagocytes and platelets in situ. The chemotactic factor directs further phagocytes to the site of the reaction and the ROIs produced aggravate the endothelial damage. (After Frank and Massaro [175])

of the pulmonary vessel walls is the most sensitive to oxygen toxicity, since it is the target of attacks from two directions (Fig. 25): from increased intracellular $O_2^-\cdot$ production in endothelial cells and from the conversion of a plasma factor to a chemotactic factor under the effect of $O_2^-\cdot$ released by phagocytes and platelets in situ (see Sect. 2). The chemotactic factor attracts further phagocytes which release additional ROIs, while platelets attached to the damaged endothelium generate $O_2^-\cdot$leading to even more severe damage.

Tolerance to oxygen toxicity can be produced in experimental animals by exposure to sublethal hyperoxia. This induces the defence systems (SOD, GP, CAT), and the increased enzyme levels to fight off subsequent hyperoxia, which is lethal in controls. The free radical theory has been proved by the successful treatment of the symptoms of oxygen toxicity with antioxidants (vitamin E, SOD, D-penicillamine), e.g. in neonatal diseases caused by oxygen therapy (retrolental fibroplasia, bronchopulmonary dysplasia) (see Sect. 20). In experimental animals survival increased significantly and pathological changes in the lungs decreased if endotoxin had been administered before hyperoxic exposure. The protective effect of endotoxin has been shown to be due to its increasing the activity of pulmonary antioxidant enzymes. The survival of rats exposed to 100% oxygen inhalation increased significantly in response to administration of CAT and SOD jointly enclosed in a liposome, as opposed to intravenous SOD and CAT, which were ineffective. The liposome-enclosed enzymes strongly increased the pulmonary SOD and CAT activities, hence their protective effect. The absence of effect upon intravenous administration of the enzymes was probably due to their short half-lives and their inability to enter the cells because of their high molecular weight [107, 117, 174, 175, 529, 559].

1.1 Summary

On the molecular level oxygen toxicity is due to the reaction of oxygen free radicals with cellular components, In hyperoxia, the intracellular generation of free radicals increases and cell damage occurs owing to the breakdown of the defence mechanisms.

2 The Role of Free Radical Reactions in Inflammation

Part of the $O_2^-\cdot$ and H_2O_2 generated when phagocytes are activated leave the cells. Antioxidant protection is weak in the extracellular space and thus the intermediates released may cause tissue damage [467, 478]. It has been shown that $O_2^-\cdot$ reacts with a plasma precursor resulting in neutrophil chemotactic activity in the plasma. It has been suggested that the precursor is an albumin-lipid complex, which is probably released by activated phagocytes [147, 177, 357, 359, 432]. In addition, PG synthesis is initiated by the activation of membrane phospholipase A_2 and some of the prostaglandins also enhance inflammation. 5-HETE and LTB_4 produced in the lipoxygenase pathway also have chemotactic effects (their effect corresponds in strength to that of C5a), while

SRS-A also increases microvascular permeability in addition to its characteristic effect on the bronchi [204, 479].

Free radicals have been shown to inactivate the normal serum inhibitors of leukocyte proteases. For instance, free radicals inactivate α 1-antiprotease, the inhibitor of leukocyte elastase, and this may play a role in lung damage caused by ROIs [60, 64, 148, 185, 278]. The chemotactic factor only delivers further neutrophils to the site of inflammation, thus establishing contact between inactive and activated cells, but does not activate the respiratory burst or cause degranulation. This is important for the termination of the process, because if for example the superoxide-dependent plasma chemotactic factor activated the respiratory burst, more $O_2^{-}\cdot$ would be released, and the process could not be terminated. As long as phagocyte activating substances are found at the site of inflammation, the process is stimulated by ROIs released from the accumulated leukocytes. These ROIs attack the lymphocytes, leading to the production of further intermediates and lysosomal enzymes. Chemotactic factor is also produced, leading to further severe tissue damage. If the activating substance is decomposed by the phagocytes, there is no further ROI and chemotactic factor production and the process is terminated. SOD has been found to be an effective anti-inflammatory agent in several animal models. Its effect has been shown to be due to inhibiton of superoxide-dependent chemotactic activity. LTB_4 facilitates the adhesion of granulocytes to the endothelium, which is thus directly damaged by SRS-A, while its permeability increases [147, 204, 357, 359].

There is an antioxidant component in the effect of the known anti-inflammatory agents. Nonsteroid anti-inflammatory agents act not only by inhibiting cyclooxygenase, but also by the $O_2^{-}\cdot$ dismutating effect of their copper(II) complex, which is comparable in strength to that of SOD. Such an effect has been demonstrated for penicillamine, indomethacin, piroxicam and acetylsalicylic acid. The $O_2^{-}\cdot$ dismutating effect of the copper complexes is probably catalytic, as opposed to the stoichiometric effect of scavengers. Therefore, the complexes are effective in lower doses and have a more lasting effect. Compared with SOD, they also have the advantage of a low molecular weight, thus being able to enter the cells more easily than the high molecular weight SOD, which is probably active only in the extracellular space. Free, hydrated cupric (Cu^{2+}) ions have been found to have a strong SOD-like effect comparable to that of the copper complexes, or even stronger [34, 319, 466, 514]. Several nonsteroid anti-inflammatory agents have been shown to inhibit the respiratory burst of phagocytes. Other authors, however, described an opposite, stimulating effect [515, 528]. The anti-inflammatory effect of gold compounds has not so far been fully explained. Besides other mechanisms (inhibition of lysosomal release, moderate inhibition of PG synthesis), the inhibition of respiratory burst by gold compounds has been suggested and lately also proved [464, 528]. Glucocorticoids, too, inhibit the respiratory burst, which effect may be related to their inhibitory effect on phospholipase A_2 activation [520].

Two kinds of reaction are distinguished in the phagocyte-target cell interaction. One is the “innocent bystander” reaction. In this reaction toxins are released by the stimulated phagocytes, which attack the surrounding cells nonspecifically. In the other type of reaction, the phagocytes attack their antibody-coated target cells specifically. This is called an antibody-dependent cellular cytotoxicity (ADCC) reaction. Red blood cell damage in which $O_2^-\cdot$ is involved belongs to the first type. It has been shown experimentally that $O_2^-\cdot$ damages the cell interior, and not the outer membrane, because it enters the cell through anion channels and oxidizes haemoglobin to methaemo globin

$$\underset{\text{Oxy-Hb}}{O_2 + \text{Hb-Fe}^{2+} \ldots O_2} \xrightarrow{2H_2} \underset{\text{Met-Hb}}{H_2O_2 + \text{Hb-Fe}^{3+}} + O_2$$

$$H_2O_2 + \text{Hb-Fe}^{3+} \longrightarrow (\text{Hb-Fe}^{3+} \ldots H_2O_2) \rightarrow \text{Lysis}$$

The H_2O_2 produced forms a peroxide-methaemoglobin complex, which has a strong oxidizing potential. This eventually leads to cell lysis [147, 498, 573].

Of the intermediates produced by phagocytes, H_2O_2 has been shown to damage lymphocyte function and to cause changes in the cell membrane [308]. According to other authors, H_2O_2 stimulates in vitro lymphocyte mitogenesis, while in higher concentrations it has an inhibitory effect. Thus H_2O_2 is characterized by a lymphocyte function modulating action [466]. PG synthesis also plays a role in the regulation of lymphocyte function, and free radicals may influence lymphocytes through this mechanism [324]. In haemodialysis, complement, activated by the alternative pathway, causes pulmonary leukostasis, leading to pulmonary arterial hypertension and oedema (for the mechanism see Sects. 7 and 14). Toxic ROIs, which are released from the complement-activated granulocytes and damage the pulmonary endothelium, play a part in this process [183, 475]. ROIs also play a role in the killing of tumour cells by phagocytes [81, 573].

The role of ROIs is not clear in the second type (ADCC) phagocyte-target cell interaction. They probably participate in the killing of the antibody-coated target cells. This reaction is rather difficult to study because the antibodies adhering to the cell membrane make it difficult for scavengers to reach th site of ROI production. The oxidative mechanism of phagocytes may be activated in the ADCC reaction, but to a lesser extent than in nonimmune phagocytosis. Owing to the close effector-target contact, however, oxygen radicals may have also a damaging effect in much lower concentrations. The increased phagocytosis of opsonized particles and immunoglobulin-sensitized cells is well known. They activate the respiratory burst by binding to the Fc and/or C3b receptors of phagocytes. This binding of sensitized cells is important in their disappearance from circulation, and it can be postulated that the ROI-induced damage of sensitized cells and particles plays an important role in cell and tissue damage in some autoimmune diseases [206, 573].

2.1 Summary

Part of the ROIs generated in activated phagocytes leave the cell and enter the extracellular space. Since antioxidant protection is weak here, the radicals may damage healthy cells and tissues. It has been postulated that $O_2^-\cdot$ reacts with a plasma *precursor, leading to chemotactic activity in the cytoplasm. Free radical reactions participate in inflammatory processes by inducing chemotactic activity, by virtue of their relation with arachidonic acid metabolism and by inhibiting the serum inhibitor of leukocytic proteases. SOD inhibits inflammation by interfering with the $O_2^{-}\cdot$-dependent chemotactic activity. ROIs are involved in both types of phagocyte-target cell interactions (bystander and ADCC), although their role in the latter type is not clear. Inhibition of pathological free radical reactions has been supposed to account for part of the action of the known anti-inflammatory agents. Glucocorticoids inhibit the phospholipase A_2-mediated activation of arachidonic acid metabolism and, in addition, or consequently, the respiratory burst of phagocytes. Nonsteroid anti-inflammatory agents probably act not only by inhibiting cyclooxygenase activity, but also by the SOD-like $O_2^{-}\cdot$ dismutating action of their copper complexes and by inhibiting the respiratory burst of phagocytes. The mechanism of the anti-inflammatory effect of gold compounds is not known, they probably inhibit the respiratory burst of phagocytes.*

3 Relation of Free Radical Reactions to the Immune System and Their Role in Immunopathological Syndromes

There seems to be an intricate set of relations between free radical reactions and the immune system. Various immune reactants (immunoglobulins, immune complexes, complement components and lymphokines) may activate the respiratory burst. This may occur in isolation, i.e. may not be associated with degranulation, cell migration and activation of phagocytes [184, 206, 286, 394]. ROIs (e.g. H_2O_2 see Sect. 2) are able to modulate lymphocyte activity. $O_2^{-}\cdot$ inhibits T-lymphocyte activity, therefore SOD activation is necessary for the normal function of T-lymphocytes [236, 308]. ROIs have also been supposed to be involved in the mechanism of action of the factor inhibiting the migration of macrophages (MIF). MIF probably activates the respiratory burst of macrophages and the reversible autotoxicity of the ROIs produced (e.g. HOCl generated in the H_2O_2-MPO-halide system) may be responsible for the inhibitory effect. A similar phenomenon has been described: ROIs formed at the activation of the respiratory burst of neutrophilic leukocytes inhibit the spontaneous and chemotactic migration of the cells [277]. Modulation of arachidonic acid metabolism by free radical reactions is an important link with the immune system, since PGs, the end products of this metabolic chain, and the products of the lipoxygenase pathway are potent immunomodulators.

The mechanism suggested by Craddock et al. to explain the transient leukopenia (granulocytopenia and monocytopenia) plus respiratory functional disturbances occurring in patients under haemodialysis care is a good example of the connection between the two systems. Similar mechanisms are probably active in a number of other diseases (Fig. 26). The cellophane membrane used in haemodialysis causes complement activation by an alternative pathway. C5a is formed, which triggers the events shown in Fig. 26. C5a is also a known activator of the respiratory burst, and the intermediates released initiate the

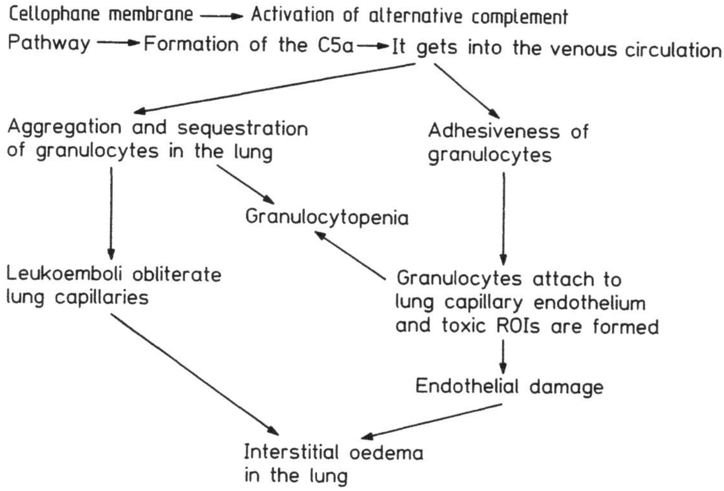

Fig. 26. Postulated mechanism of leukopenia and respiratory functional disturbances occurring during haemodialysis. (After Füst [183])

formation of the superoxide-dependent chemotactic factor as described and the production of LTB_4 and SRS-A owing to the connection with arachidonic acid metabolism. These substances cause a further increase in granulocyte sequestration and play a role in the adhesion of granulocytes to the vascular wall and in endothelial damage. A similar mechanism may play a role in other diseases with the difference that in those the process is initiated by some other factor (immune complex, endotoxin, etc.). These factors may be responsible for phenomena such as leukopenia in autoimmune diseases (systemic lupus erythematosus, Felty's syndrome, rheumatoid arthritis), for the transient neutropenia in donors during filtration leukopheresis, and for the symptoms of acute respiratory distress of adults occurring in trauma, Gram-negative sepsis or acute pancreatitis [90, 110, 183, 275, 475]. In autoimmune disease the circulating immune complexes activate the respiratory burst of granulocytes, and thus increased amounts of ROIs enter the extracellular space continuously. The arterial activation of the Craddock mechanism may occur in vivo and this may account for the pathomechanism of autoimmune vasculitis. These theories have been experimentally verified [183, 235, 239, 521].

Chromosome instability (fragility) has been found in several autoimmune diseases (systemic lupus erythematosus, rheumatoid arthritis, scleroderma, dermatomyositis, periarteritis nodosa, ulcerative colitis, Crohn's disease). This sign could be detected in all of these diseases except rheumatoid arthritis (RA), where it occurred in only one-third of the cases. A clastogenic factor responsible for chromosomal fragility was found in the serum of scleroderma, systemic Lupus erythematosus patients and of those RA patients who showed increased fragility of the chromosomes. The exact nature of this factor is not known, it is probably a low molecular weight substance derived from monocytes. The joint occurrence of chromosomal fragility, abnormal immune response, and increased frequency of tumours has been known for some time, but the connection between them was not clear. This triad has been described in autoimmune diseases, in some hereditary diseases with DNA repair deficiency (Bloom's syndrome, Werner's syndrome, Fanconi's anaemia, ataxia telangiectasia, Down's syndrome) [136, 393, 404]. A decreased SOD content of red blood cells has been found in Fanconi's anaemia [283], while in Down's syndrome the level of mitochondrial MnSOD decreases, while the cytoplasmic CuZnSOD level increases owing to the gene dose effect (CuZnSOD is coded for by chromosome 21) [53, 399, 404, 407]. A decreased total SOD level of the granulocytes was also found in juvenile RA patients [458, 459]. In adult RA the total SOD level did not differ significantly from the control value, but the cytoplasmic MnSOD level of granulocytes was significantly lower [423]. The fibroblasts of Bloom's syndrome patients also produce the clastogenic factor spontaneously. According to Cerutti, Bloom's syndrome is a chronic promotion state in which carcinogenesis has been initiated [134, 135, 602]. The triad also occurs in New Zealand Black (NZB) mice, regarded as the animal model of autoimmune disease. As a result of the decrease in suppressor T-cell activity, autoimmune disease develops spontaneously in these mice in association with Coombs-positive haemolytic anaemia, immune complex glomerulonephritis and increased production of antinuclear antibody [136, 235].

At the same time, treatment of RA and Crohn's disease patients with D-penicillamine-copper complex has recently been shown to be able to dismutate $O_2^-\cdot$, which is a SOD-like action. SOD decreases chromosome fragility in the blood culture of patients with Werner's syndrome and Fanconi's anaemia as well as other autoimmune diseases [136, 372, 404].

The connection between the members of the triad may be an increased $O_2^-\cdot$ flux (due for example to increased $O_2^-\cdot$ production caused by ADCC or by some protective enzymes; see also Sects. 10 and 11). The deficiency of DNA repair observed in hereditary diseases may be secondary: the repair mechanism becomes insufficient in the face of gradually increasing DNA defects. This may be responsible also for the anti-DNA antibody production in autoimmune diseases, as evidenced by their occurrence in xeroderma pigmentosum caused by DNA repair deficiency [136].

Another interesting connection has been found between connective tissue diseases and free radical reactions. The enzyme prolyl hydroxylase, which uses O_2 for the conversion of some proline components of protocollagen to hydroxyproline in the presence of iron, α-ketoglutarate and vitamin C, probably uses oxygen radicals for its action. This step is needed for the formation of hydrogen bonds between the polypeptide chains and for the development of the triple helix structure, which enables the molecule to leave the fibroblast. This is a possible mechanism of the fibrosis-inducing effect of free radical reactions, proved by the fact that some drugs with known oxygen radical-releasing effects (bleomycin, nitrofurantoin) and also irradiation cause pulmonary fibrosis [43]. In rats chronic exposure to bleomycin caused skin changes which closely resembled the changes occurring in scleroderma in humans as regards their clinical, histological and ultrastructural characteristics. It has been suggested that this rat experiment could be used for modelling human scleroderma [383].

3.1 Rheumatoid Arthritis

It is for the free radical pathomechanism of RA that the highest number of data are available. RA is an immune complex (IC) disease; the joints of the patients contain all components of a local Arthus reaction (white blood cells, IC, complement). ICs and complement (C) deposited in the synovium and articular cartilage activate the respiratory burst of the neutrophils present and cause degranulation. The process described in the chapter dealing with inflammation is initiated and perpetuates the inflammation (there is abundant experimental evidence for LPO damage occurring in the joints of RA patients) [24, 470, 521, 577].

During its redistribution, iron, being stored as ferritin in the synovium in addition to the usual sites, is transformed from Fe(II) to Fe(III) and may generate ROIs by reducing molecular oxygen. As a result of iron surplus, ferritin becomes saturated and loses its antioxidant effect, moreover it may act as a prooxidant. The saturated ferritin is concentrated in the lysosomes of the synovial reticuloendothelial system (RES) cells and enhances inflammation through the peroxidative damage of the lysosomes. Iron thus plays a part in the acute phase of inflammation. In addition, it also participates in the transport of the infiltrate composed of lymphocytes and macrophages in the acute phase, probably using the iron-binding proteins on the surface of these cells [41]. As stated, the decrease of total SOD activity has been demonstrated in the neutrophils of children with juvenile RA. In adults with RA the decrease of cytoplasmic MnSOD activity has been found which leads to the growth of extracellular $O_2^- \cdot$ flux [458, 459]. Direct proof has been obtained as regards the damaging effect of free radicals on cartilage: significant inhibition of proteoglycan synthesis has been found in cultures of intact bovine articular cartilage in response to $O_2^- \cdot$ produced enzymatically in the xanthine oxidase-hypoxanthine system. Similar changes have been observed in RA with cartilage biopsy made at an early stage, i.e. soon after the appearance of the symptoms [27].

Convincing evidence in support of the free radical pathogenesis of RA is provided by the excellent effect of intra-articularly administered SOD. The early effect of SOD is comparable to that of corticosteroids, while its late effect is even better. Moreover, it has no significant side effects [202]. Partial success can be obtained in systemic lupus erythematosus (SLE), dermatomyositis and Crohn's disease with intramuscular SOD; the treatment decreases chromosomal fragility and the level of antinuclear factor [136, 404].

3.2 Systemic Lupus Erythematodes

As has been shown, free radical reactions most probably play a role in the development of some of the symptoms of SLE (leukopenia, vasculitis). Harman has provided additional evidence for the primary causative effect of these reactions in the development of SLE. In NZB mice (the animal model of SLE) antioxidants added to the diet increased the average life span of the animals. The quinoline derivative Santoquine (generic name, ethoxyquin) increased the mean survival by 32.1%. Other authors described similar results in NZB/NZW mice, the F_1 hybrids of NZB mice, with the use of the disodium salt of the water-soluble *N*-(Carboxyphenyl)-4-Chloroanthranylic acid (CCA), an antioxidant with immune modulator effect. It is also effective in RA. In mice CCA inhibited the development of autoimmune nephropathy, and decreased the serum autoantibody level, probably by increasing the activity of suppressor T-lymphocytes. According to Harman, the autoimmune manifestations of SLE resemble the autoimmune phenomena of ageing, with the difference that those of SLE appear earlier and are more pronounced. The free radical reactions asociated with ageing (see Sect. 4) decrease both the humoral and the cellular immune response, but by inhibiting the function of suppressor T-cells, which actively maintain tolerance to the body's own antigens, increase autoimmune reactions in both animals and humans. T-cells are relatively radioresistant, but their precursors are radiosensitive, i.e. also sensitive to free radical reactions.

These results have led to the formulation of the following hypothesis concerning the pathogenesis of SLE. The basic, inherited defect (or defects) in SLE increase antigen formation from nuclear components by free radical reactions, which are accelerated by some factor. The cell and tissue damage caused by free radical reactions may increase the antigenicity of some cell components. Administration of halothane and CCl_4 has been found to increase the formation of autoantibodies and immune complexes. If this hypothesis is true, the exacerbations and remissions characterizing SLE must be strongly influenced by factors decreasing or increasing endogenous free radical reactions. Several observations have substantiated this theory.

q) 1. Exposure to sunlight has an unfavourable effect in SLE; UV light is known to induce free radical reactions also involving nuclear components.

q) 2. SLE is frequent in women of reproductive age in whom the level of serum copper, a known catalyst of free radical reactions, is higher than in men. Oral contraceptives increasing the serum copper and serum lipid levels may cause the appearance of LE cells and the exacerbation of SLE.

q)3. Several of the SLE-inducing agents are known to accelerate free radical formation, e.g. isoniazid (INH) and methyldopa produce free radicals during their cytochrome P-450-mediated metabolism. Hydralazine, a hydrazine derivate like INH, has a similar effect.

The free radical reactions supposed to affect nuclear components may induce the expression of type C viruses observed in SLE. χ-irradiation, methylcholanthrene and 7,11-dimethylbenz(α)anthracene can activate genetically transmitted viral genomes and thus lead to the appearance of type C viruses. The cell types from which the nuclear antigens originate are absent in SLE; most probably the suppressor T-cell precursors sensitive to free radical reactions and fibroblasts play a role in this disease. Finally, of the agents causing remission, chloroquine and steroids inhibit the activation of the arachidonic acid cascade by free radical reactions. The favourable effect of Levamisole is probably due to its antioxidant action (see Chap. VII Sect. 2). Chloroquine may inhibit free radical reactions by the abstraction of the hydrogen from the nitrogen bonded to its pyridine ring [19, 235, 243, 246].

3.3 Summary

Free radical reactions have a complex relation with the immune system. Various immune reactants (immunoglobulins, immune complexes, complement, lymphokines) may activate the respiratory burst of phagocytes. ROIs are able to modulate the function of lymphocytes. Immune substances may activate the respiratory burst not only in experiments, but also in various human diseases, e.g. in leukopenia and disturbed respiratory function during haemodialysis (the cellophane membrane activates the complement system by an alternative pathway) and in leukopenia and vasculitis occurring in autoimmune diseases (e.g. SLE, Felty's syndrome and rheumatoid arthritis). The triad of chromosome fragility, abnormal immune response and increased tumour frequency is a common characteristic of most autoimmune diseases and of syndromes associated with deficient DNA repair. The common denominator of the members of the triad is probably an increased $O_2^{-}\cdot$ *flux. This is still a hypothesis, but is supported by experimental evidence. The involvement of free radical reactions in the development and in the symptoms of connective tissue diseases (leukopenia, vasculitis, arthritis, fibrosis) has been proved, first of all in rheumatoid arthritis, but it has also been suggested that they play a primary role in the pathogenesis of RA or SLE for example. This, however, still needs verification.*

4 The Free Radical Theory of Ageing

Although ageing is a physiological process and thus it would be more appropriate to discuss its relation with free radical reactions in the chapter dealing with physiological free radicals, its close association with other, pathological processes caused by free radicals justifies its inclusion in this chapter. Two factors play a major role in ageing: (a) the genetically determined ageing process, which affects all individuals of a species, independent of the environment and (b) accidental ageing, which is due to damaging environmental effects, e.g. diseases.

Life expectancy at birth is the number of years an individual may reasonably expect to live under the specific mortality conditions of the species. This value is a statistical average and is not identical with the highest age the individual may reach. In most of Europe, life expectancy is 68-72 years, and in the developed countries this value is increasing owing to the high level of medical care and public health. The average life expectancy around 70 years of the population of the developed countries is still 15 years less than the 85 years that healthy individuals may expect to live. Potential life expectancy is the maximum number of years that an individual of a species can be expected to live as long as the genetic characteristics of the species remain unchanged. The potential life expectancy of humans is about 100 years, but a maximum of 120 years has also been suggested. The 15 years or so difference between the maximum life expectancy and the figure of about 85 years that healthy individuals may expect to live is due to age-independent accidental deaths and to individual differences which may be observed even in the absence of disease [31, 229, 232, 233].

It has been shown in animal experiments that administration of antioxidants and LPO-decreasing diets containing little copper, little unsaturated fatty acids and much natural antioxidants (selenium, vitamin E) and essential nutrients increases the average life expectancy of the animals. These measures to minimize free radical reactions will, however, in most cases increase only the average life expectancy and will not influence the potential life expectancy, which was found to change in a few cases only [229, 231, 232, 233, 246, 491].

The potential life expectancy of some species shows a close correlation with the animals' basal metabolism (O_2 uptake per unit body weight per day). In most mammals the two parameters are inversely related, i.e. their product, called the life span energy, is constant. Its value is 220 kcal/g in most mammals. In some species, e.g. in the cat and some nonhuman primates, it is higher, approx, imately 558 kcal/g. It is highest in humans in lemurs and in the weeper monkey: around 781 kcal/g. These data show that the maximum life expectancy of the species depends on their oxygen uptake [102].

According to Harman, ageing is due to the cumulative damaging effect of "univalent leakage" on mitochondria. Some 90% of the O_2 uptake is used in the mitochondria. Probably the rate of O_2 uptake is under genetic control, in which both nuclear and mitochondrial genetic material is involved. $O_2^-\cdot$, small amounts of which are constantly produced in the respiratory metabolic chain in proportion to the basal metabolism, causes cumulative damage in the mitochondria, because the efficiency of the protective systems decreases with age. The rate of extramitochondrial damage is also increasing. Morphological changes in the mitochondria and a decrease in the activity of mitochondrial enzymes have been demonstrated in relation to ageing. Thus the mitochondrium may be regarded as a kind of biological clock, and death occurs when mitochondrial function falls below a critical level in the entire organism or in organs of vital importance, like the central nervous system. Antioxidants or diets reducing LPO to a minimum seem to decrease only extramitochondrial free radical reactions and thus increase only the average life expectancy, not the maximum life span. There are, however, methods available whereby the potential life expectancy can be influenced, namely severe caloric restriction and the lowering of the body's surface temperature.

The free radical theory of ageing may explain the success of these measures, which are suitable for influencing the basal metabolism and thus univalent leakage proportional to it. Restriction of caloric intake decreases the amount of reducing equivalents transported to the mitochondrial respiratory chain in the Szent-Györgyi-Krebs cycle and thus less oxygen is needed for water production, leading to a decreased basal metabolism [229, 232, 233, 234, 442, 484, 565]. The relation between respiration and the maximum life span is also shown by the fact that small animals with a high respiratory rate live a shorter life than large animals with low respiratory rates. In the housefly, the rate of respiration depends on the animal's activity. In an experiment, flies kept in large containers where they had room for flying had a shorter maximum life span than those kept in small containers where they could not fly [442]. Physical activity had no significant effect on the serum SOD and CAT levels, but the concentration of inorganic peroxides increased, indicating the enhancement of free radical reactions [511]. In another study of houseflies, the antioxidant defence (SOD, CAT, total glutathione level, reduced glutathione, chloroform-soluble antioxidants) were found to decrease with ageing, whereas the level of inorganic peroxides increased. The data in the literature on the changes occurring in antioxidant protection with ageing are still rather controversial [512].

There is direct experimental evidence proving that free radical reactions are more frequent in animals having a higher basal metabolism and thus (according to the free radical theory) shorter maximum life span, than in animals with lower basal metabolism and longer maximum life span. Thymine oxidation products (thymine glycol, hydroxymethyl uracil) are formed during the oxidative damage of DNA. The DNA repair mechanisms remove these products which can then be demonstrated in the urine of experimental animals and humans. The amount per Kilogram body weight of thymine oxidation products in the urine of rats having a higher basal metabolism and shorter maximum life span was 15-fold the amount measured in human urine [481].

The theory of Cutler on ageing resembles Harman's theory because he suggests that pathological free radical reactions resulting from univalent leakage which is proportional to the basal metabolism play a key role in ageing. He, however, maintains that these reactions do not cause the cumulative damage of mitochondria and/or other cell constituents, but act on the regulator genes. Thus, according to Cutler, ageing is, in fact, a process of dysdifferentiation. He thinks it probable that all mammals have identical structural genes, and the species differences are due to the function of regulator genes, which control the expression of the structural genes. Regulator genes constitute only a very small part of the genome, and it might be supposed that, among other species-specific characters, they also control the maximum life span characteristic of the species. The basal metabolism is also genetically controlled, as is probably antioxidant production. The changed function of regulator genes does not lead to the death of the cells, only to changes in their differentiation and to the inadequate expression of some of the structural genes. Changes in regulator gene function may be brought about by the normal dynamic nature of DNA, as a result of which DNA sections are interchanged or rearranged, and do not necessarily result from point mutations. Such rearrangements of DNA may be elicited even by extremely low levels of mutagens. Oxygen metabolism may produce such low mutagen levels as a result of univalent leakage. Thus, the most important effect of the free radicals present in small amounts in the normal cell may be a change in cell differentiation rather than point mutations or other types of cell damage. The cumulative damage theory of Harman is contradicted by the absence of any specific cell damage which could be held responsible for ageing or would show accumulation with ageing.

The proposition that species with higher potential life expectancy are resistant to dysdifferentiation caused by mutagenic agents has been supported by experiments in which the cells of species with a shorter life span could be more easily transformed with mutagens than could human cells. If it is true that the maximum life span is influenced by ROIs produced in the course of oxygen metabolism, the antioxidants providing protection against them must also play an important role and there must be a positive correlation between the tissue concentration of antioxidants and the maximum life expectancy. Such a correlation was, in fact, found for SOD, urate, β-carotene (the correlation was not significant for vitamin A), vitamin E, and coeruloplasmin. No correlation, or a negative correlation was found for ascorbic acid, glutathione, glutathione peroxidase, and glutathione-*S*-transferase. The higher level of some antioxidants in species with a higher maximum life expectancy may, in fact, be due to changes in regulator gene function. For instance, in humans the high level of urate is due to the absence of the enzyme uricase. Vitamin A is produced from β-carotene by the mediation of intestinal carotenase; although the carotenase activity has not been measured in the human intestine, it may be that the high carotene level in human tissues is due to a low level of this enzyme.

The antioxidants showing a negative correlation seem to be unimportant from the point of view of maximum life expectancy. Ascorbic acid is known to behave as a prooxidant in the presence of Fe^{2+}. Cytochrome P-450 and cytochrome P-448, which participate in the metabolization of xenobiotics and form ROIs in the course of their catabolic activity, show a negative correlation with maximum life expectancy.

In the studies mentioned, the tissue levels of some major antioxidants were measured, therefore the results furnish no evidence for a higher total antioxidant capacity in species with a high maximum life expectancy. Therefore, the autooxidation of tissue homogenates was measured using thiobarbituric acid reactive substances, whereby data could be obtained on the total antioxidant capacity of the tissues. The expected negative correlation was obtained for the individual species. A further difficulty, which has to be overcome when trying to prove this theory, is the variability observed in the tissue antioxidant levels in individuals of the same species, although the maximum life span is uniform within the species. Also, diets containing antioxidants failed to prolong the animals' maximum life span, in most experiments only the average life expectancy increased. Cutler [102] eliminated this possible objection by maintaining that it is not the level of the individual antioxidants which is uniform in the species, but the total antioxidant capacity. If the level of one antioxidant decreases, this may result in a compensatory increase in the level of another antioxidant. Experimental evidence has indeed been obtained for this compensatory mechanism. Thus, the total antioxidant capacity cannot be influenced by the dietary intake of one or the other antioxidant, because this may lead to a decrease in the synthesis of endogenous antioxidants. Therefore, the maximum life span of animals cannot be influenced by dietary measures. Antioxidant capacity is probably a constant value in all differentiated cells and must be sufficient for preventing dysdifferentiation.

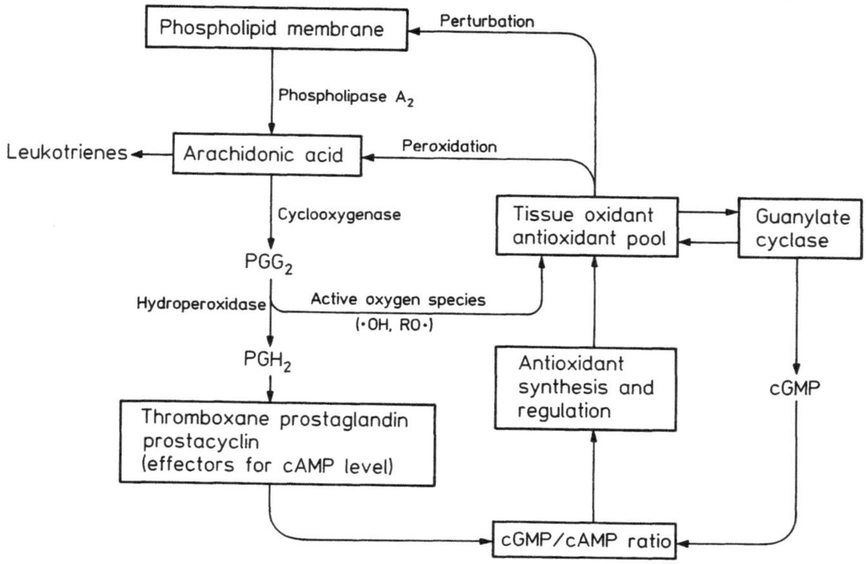

Fig. 27. Guanylate cyclase-arachidonate-cyclooxygenase model for the regulation of the redox potential of a cell, for explanation see text. (After Cutler [102])

What are the mechanisms responsible for maintaining a constant species-specific antioxidant (and thus ROI) level in the tissues? According to Cutler's model, the cells contain substances which are sensitive to the effect of ROIs. These are: (a) arachidonic acid; (b) cyclooxygenase; and (c) guanylate cyclase. ROIs enhance the release of arachidonic acid from phospholipids by activating phospholipase A_2. In higher concentrations they may peroxidize arachidonic acid. Low lipid peroxide levels activate cyclooxygenase, while high levels inhibit the enzyme. ROIs increase the synthesis of guanylate cyclase, which catalyses cGMP synthesis. The guanylate cyclase-arachidonic acid-cyclooxygenase (GAC) model is shown in Fig. 27. The three components of the system, which are sensitive to the oxidation-reduction state of the environment, may change the cGMP/cAMP ratio, thus affecting the endogenous antioxidant synthesis. Various prostaglandin derivatives may elevate the cAMP level in the cell (e.g. TXA_2, PGI_2). If ROIs are in excess in the cell, they peroxidize arachidonic acid, and there will be less arachidonic acid available for PG synthesis. Consequently, less PG will be produced, and the intracellular cAMP level will also be lower. Guanylate cyclase is activated by free radical reactions and thus the cGMP level is increased. If there is oxidant excess, the cGMP/cAMP ratio increases. Antioxidants cause an opposite shift in the ratio by eliminating these effects. The ROIs produced during the $PGG_2 \rightarrow PGH_2$ conversion during PG synthesis are responsible for an important negative feedback mechanism (in case of antioxidant excess, PG synthesis increases, the amount of ROIs released at the $PGG_2 \rightarrow PGH_2$ conversion increases, and, in turn, these decrease PG synthesis). The increase in the cGMP/cAMP ratio produced by oxidant excess acts in the direction of increased endogenous antioxidant synthesis, while the decreasing cGMP/cAMP ratio found at antioxidant excess tends to decrease antioxidant synthesis.

The validity of this model can be tested from several aspects. If it is correct, biological regulation can be cheated, e.g. by inhibiting cyclooxygenase with aspirin or indomethacin, and the resultant decrease in the PG level and, consequently, in the cAMP level simulates oxidant excess. Thus it is possible that nonsteroid anti-inflammatory agents act not only by inhibiting PG synthesis, but also by increasing endogenous antioxidant synthesis in the cells. Another important conclusion which may be drawn from this model is that large doses of antioxidants may decrease the cGMP/cAMP ratio and thereby cause a decrease in endogenous antioxidant synthesis. Thus the total antioxidant defence cannot be enhanced with exogenous antioxidants. Administration of cGMP increased the maximum life span of the nematode *Caenorhabditis elegans*. Resistance to radiation can be increased with cAMP added to cell cultures or given to mice. This theory is also supported by the results of experiments in which anaemia and thrombocytopenia of vitamin E-deficient animals could be eliminated with aspirin administration. Aspirin inhibits cyclooxygenase and thereby reduces PG synthesis, leading to an increase in the synthesis of other antioxidants which are able to compensate for vitamin E deficiency. The number of TBA reactive substances showed a 50% drop in the serum of rabbits following aspirin administration. This, however, could be due directly to the inhibition of PG synthesis and the consequent absence of ROI production [102].

Prior to the emergence of the free radical theory of ageing, the cross-linkage theory prevailed, which had been based on a large number of experimental data. According to this theory, the cross-linking of macromolecules (nucleic acids, proteins) takes place gradually, causing functional damage in the cells. It has been shown that such cross-linkage of macromolecules occurs during pathological free radical reactions [428].

Centrophenoxine (Lucidryl; generic name, meclofenoxate) used in psychiatric practice decreases the accumulation of lipofuscin in nondividing cells and has been shown to increase the average life span in experimental animals. It has been suggested that it is able to prolong the maximum life span. Lipofuscin accumulates mainly in postmitotic cells (myocytes, neurones). While decreasing the cerebral lipofuscin level, centrophenoxine has been found to improve the learning ability of animals. Dimethylaminoethanol is the active ingredient with ·OH scavenger activity. Dimethylaminoethanol is incorporated into the membrane of cerebral cells, where it is active as phosphatidyldimethylaminoethanol [208, 599].

Some of the changes and diseases associated with ageing may be explained on the basis of the free radical theory, e.g. the increased frequency of tumours, atherosclerosis and hypertension, immune dysfunction indicated by the appearance of autoantibodies and a decrease in immune function (free radicals inhibit the suppressor T-cell function and decrease both cellular and humoral immunity), changes in the central nervous system, and senile amyloidosis (see the appropriate sections). Changes in the cyclic nucleotide balance occurring with ageing may have a part in immune dysfunction: the cellular cGAMP level increases and the cAMP level decreases. Free radical reactions cause a similar shift in their ratio. The reverse is also true: accelerated ageing may be observed in diseases in which extracellular free radical flux plays a role, e.g. SLE, Down's syndrome. In addition, the same chromosome, 6, has been found to carry in humans the codes for mitochondrial MnSOD, the main histocompatibility complex (MHC) controlling immune response, the microsomal mixed function oxidases and cyclic nucleases [88, 232, 233, 236, 368, 565].

4.1 Summary

The increasing effect on average life span of antioxidants and of LPO-minimizing diets has been proved in several animal experiments. Pathological free radical reactions undoubtedly play a role in ageing, but this role may not be important and the reactions may be secondary to degradation processes elicited by the primary cause or causes. Nevertheless, it has been maintained by several authors that $O_2^{-}\cdot$ *leakage occurring during the function of the mitochondrial respiratory chain, the amount of which is proportional to the basal metabolism, may be a primary factor in ageing owing to its slow, cumulative damaging effect on mitochondria, or to its effect on regulator gene function. This explains the observation that potential life expectancy can be prolonged with caloric restriction decreasing the basal metabolism or by decreasing the surface temperature of the body. The absence of the effect of antioxidants on potential life expectancy, while they increase the average life span, may also be explained. According to one theory, antioxidants cannot reduce the mitochondrial damaging effect of univalent leakage, but they can influence extramitochondrial damage. Another theory maintains that the maximum life span is determined by the total antioxidant capacity, which cannot by changed by the administration of either one or several antioxidants, the more so, because antioxidant administration may lead to a decrease in endogenous antioxidant synthesis. Accelerated ageing has been observed in diseases in which there is an increased free radical flux, e.g. SLE or Down's syndrome. Although the causative role of free radical reactions in ageing has not been sufficiently proved, their involvement in the process must be taken into consideration by all modern theories proposed to explain ageing.*

5 The Role of Free Radical Reactions in the Pathogenesis of Some Neurological Diseases

5.1 Chronic Organic Brain Syndrome

In chronic organic brain syndrome (COBS) the central nervous system, or at least the part controlling higher nervous function, shows premature ageing, which is asynchronous with the process of ageing in other organs. Severe mental and emotional deterioration may be observed, while the function of other organs is still adequate. The role of free radical reactions in COBS may be supposed on the basis of experiments in which the maze learning ability and Skinner box learning of rats decreased significantly with the elevation of the unsaturated fatty acid content of the diet and with the degree of unsaturation. Diet did not significantly influence mortality. This is in accord with the proposition that increased LPO damages the central nervous system, while its effect on other organs is less significant. This is probably due to the fact that neurones are fixed postmitotic cells having exclusive connections with other neurones. In addition, as opposed to parenchymal cells, neurones are supplied not only by the capillaries, but also by glial cells attached to the capillary wall. LPO in the central nervous system is also enhanced by the high PUFA content of synaptic regions [230, 232, 233].

Eddy and Harman have demonstrated the accumulation of unsaturated docosahexanoic acid (22:6ω3), which is incorporated in the phospholipids of neurones both in the perikaryon and in the synaptic region. The decosahexanoic acid occurring in the diet either freely or in triglyceride form, and linoleic acid, its precursor, are selectively accumulated in the brain because they probably have a role in the regulation of membrane permeability and membrane-bound enzyme activity. Being unsaturated, however, they increase the sensitivity to peroxidative damage, therefore the excessive intake of these substances in the diet may have an untoward effect [50, 129]. Chia and co-workers [77] have shown that the level of MDA and conjugated dienes increases in the human brain with ageing, indicating increased LPO.

5.2 Alzheimer's Disease

The role of free radical reactions has been suggested in the pathogenesis of Alzheimer-type presenile dementia. The neuronal accumulation of lipofuscin has been demonstrated in this disease. Senile plaques also appear in normal senescence, but their number is higher in Alzheimer's disease. The first changes corresponding to LPO-induced mitochondrial damage appear in the mitochondria of axon terminals. Neurofibrillar degeneration in the cytoplasm of neurones may be observed not only in Alzheimer's disease, but also in other diseases associated with free radical reactions, e.g. Parkinson's disease and Down's syndrome. An interesting connection has been observed between Alzheimer's disease and Down's syndrome. After the age of 20 years, lesions resembling those of Alzheimer's disease appear in patients with Down's syndrome (neurofibrillar degeneration and senile plaques). Down's syndrome is more frequent in families in which Alzheimer's disease occurs. Also, Alzheimer's disease is a frequent cause of COBS [232, 233, 234, 312].

5.3 Parkinson's Disease

In the following, a model will be described, in which manganese, neuromelanin, catecholamines and free radical reactions have been shown to play a role in the pathomechanism of parkinsonism. For a better understanding of the model it should be known that the symptoms of manganese intoxication resemble those of parkinsonism, but are not entirely the same. The early symptoms of manganese intoxication mimic psychosis (psychomotor excitation, hallucination). The late phase is characterized by permanent extrapyramidal neurological symptoms. Chronic manganese intoxication is only suitable for modelling postencephalitic parkinsonism, because it differs from idiopathic parkinsonism in several important aspects. For instance, chronic manganese intoxication is characterized by hyperkinesia and not by hypokinesia. The sites of the typical lesions are also different: in parkinsonism depigmentation and cell death occur in the substantia nigra, locus coeruleus and in the nucleus dorsalis of the vagus nerve, while the striatum and pallidum are less involved or intact. The nervous structures damaged in manganese intoxication include the pallidum, the subthalamic and caudate nuclei, and the putamen, although the substantia nigra may also be affected. Attempts to look for an increased manganes level in the basal ganglia in parkinsonism have been unsuccessful. Nevertheless, we cannot completely dismiss the idea of a connection between these two conditions. Although the high concentration of manganese in the striatum causes hyperkinetic extrapyramidal dyskinesia, the similarities are also impressive: intoxication caused by chronic exposure to small amounts of manganese is characterized by bradykinesia, low dopamine level in the striatum, and depigmentation of the substantia nigra.

The cerebral structures affected in parkinsonism, i.e. the substantia nigra, locus coeruleus, and the dorsal nucleus of the vagus, are pigmented and contain neuromelanin under normal conditions. The cytoplasmic deposition of melanin is a general characteristic of catecholamine neurones. As opposed to eumelanin, found in the skin, neuromelanin is deposited with lipofuscin and in the cytoplasm, not in the melanosomes. A further difference is that neuromelanin is formed by polymerization from the autooxidation products of dopamine, its precursor, while eumelanin is formed from the autooxidation products of L-dopa. The reason for this difference is that eumelanin production is catalysed by tyrosinase up to the dopaquinone step. This enzyme also has tyrosine hydroxylase and dopa oxidase activity, while the catecholamine neurones contain only tyrosine hydroxylase, which has no dopa oxidase activity. Therefore, L-dopa is converted to dopaminase by L-dopa decarboxylase and neuromelanin is formed as the result of the autooxidation of dopamine. Despite the structural differences, neuromelanin resembles eumelanin in function: it is a free radical scavenger, while it is formed as a result of free radical reaction, as suggested by its deposition in association with lipofuscin in the cytoplasm of catecholamine neurones. What is the role of neuromelanin which, unlike lipofuscin, has a useful function? Catecholamines (e.g. dopa, dopamine, noradrenaline, adrenaline) are each capable of autooxidation in the catecholamine neurones, in the course of which ROIs or reactive cytotoxic semiquinones and oxidation products (e.g. 6-OHDA) are formed. Catecholamines are catabolized mainly by catechol-oxy-methyl-transferase (COMT) and monoamino-oxidase (MAO), and H_2O_2 is formed during MAO activity. Neuromelanin has the function of protecting the neurone against the damaging effect of these reactive products.

As to the role of manganese, it may have a significant antioxidant or prooxidant effect, depending on its valency. Mn^{2+} is an antioxidant, while Mn^{3+} and Mn^{4+} are prooxidants. In vitro the higher valency manganese increases dopamine autooxidation and thus the formation of ROIs, cytotoxic semiquinones and oxidation products. The antioxidant effect of Mn^{2+} is due to its $O_2^{-}\cdot$ and H_2O_2 scavenger effect

$$2\,H^+ + O_2^{-}\cdot + Mn^{2+} \rightarrow H_2O_2 + Mn^{3+}$$
$$H_2O_2 + Mn^{2+} \rightarrow OH \;\; + OH^- + Mn^{3+}$$

The symptoms of manganese intoxication indicate specific lesions of some cerebral structures and not diffuse damage. These structures have a high oxidative status and the neurotransmission they participate in is sensitive to the modulating effect of manganese. In manganese intoxication the highest manganese concentration is found in the hypothalamus. The level of dopamine β-hydroxylase (DBH) enzyme is also extremely high in the hypothalamus. Ascorbic acid is the cofactor of this enzyme, it is activated by the ROIs formed during the autooxidation of ascorbic acid. Even under normal conditions the hypothalamus contains the highest Mn level of all cerebral tissues. Since there is hardly any CAT activity in the brain, and also the GP activity is very low, manganese must be one of the most important physiological scavengers in brain tissue. There is a delicate balance between manganese and the ROIs produced during the autooxidation of ascorbic acid, which are needed for the activation of DBH. Ascorbinic acid provides for the quantity and quality of ROIs needed for DBH activity, and manganese inhibits the excessive production of ROIs, which would be harmful; e.g. H_2O_2, which may inactivate DBH and ROIs that may damage the surrounding cells and tissues. This delicate balance may be upset by exogenous Mn^{2+} which, by its scavenger action may inhibit the production of ROIs needed for DBH activation. Alternatively, copper accumulation may upset the ROI-antioxidant balance leading to an excess of ROIs. Therfore, in acute manganese intoxication the hypothalamic noradrenaline level decreases if there is a drop in DBH activity.

The striatum is another structure sensitive to manganese, which causes disturbances in cholinergic and dopaminergic neurotransmission (see below). Finally the sensitivity of the substantia nigra may be explained by the high MAO activity in this region, as a result of which large amounts of H_2O_2 are produced. Because of the low CAT and GP levels, H_2O_2 reacts with manganese, which is converted to its higher valency form and enhances the autooxidation of catecholamines and thus also the formation of ROIs, which damage the substantia nigra.

Psychosis in acute manganese intoxication may thus be due to the free radical scavenger action of Mn^{2+}, causing the dysfunction of the free radical activated enzymes which participate in the synthesis and/or modulation of neurotransmitters. The extrapyramidal neurological symptoms of manganese intoxication may be due to the conversion of Mn^{2+} to Mn^{3+} and Mn^{4+} stimulating catecholamine autooxidation and ROI formation. (This also explains why in chronic manganese intoxication there is no correlation between the tissue levels of manganese and the extrapyramidal symptoms. Following acute exposure there is no need for the maintenance of high tissue manganese levels for the development of neurological symptoms.) ROI production caused by high-valency manganese leads to the gradual destruction of the catecholamine-containing neurones, as a result of which the dopamine level may decrease in the striatum and depigmentation occurs in the substantia nigra. The drop in the dopamine level in the striatum is due to damage to the substantia nigra and to the dopaminergic nigrostriatal neurones. The damage is caused by the decrease of the free radical scavenging neuromelanin in the substantia nigra. Manganese, similarly to neuromelanin, may thus be regarded as an important free radical scavenger of the central nervous system. It plays a major role in the control of neurotransmission and in the protection of the central nervous system.

Depigmentation of the substantia nigra and a low striatal dopamine level are also found in Parkinson's disease. According to Barbeau, idiopathic parkinsonism is triggered by increased dopamine turnover, which may be elicited by various events. Increased dopamine turnover leads to an increase in dopamine autooxidation and ROI formation, exceeding the scavenger activity of neuromelanin and other antioxidants and causing cell destruction. The striatal dopamine level decreases owing to the damage of nigrostriatal neurones. An increased dopamine turnover may be produced by any action blocking dopaminergic neurotransmission in the synapse or disturbing the feedback mechanism. The blockade is compensated for by increased dopamine synthesis, and thus autooxidation also increases. Haloperidol for example acts postsynaptically: it prevents dopamine from reaching the receptors. Manganese increases dopamine autooxidation by being converted to a higher valency form or by inhibiting the synthesis of cAMP, the second messenger of neurotransmission. Amphetamine depletes dopamine. The reason for the increased dopamine turnover in Parkinson's disease is not known. Familial juvenile parkinsonism is probably caused by congenital neuromelanin deficiency in the substantia nigra. The death of dopamine-producing cells enhances dopamine synthesis in the surviving cells, in which there will be an increased dopamine turnover, leading to the progression of the disease [9, 21, 22, 25, 84, 123, 209, 404].

5.4 The Role of Free Radicals in Neurotransmission

Experimental evidence has shown that lipid peroxide free radicals modulate neurotransmission under physiological conditions. Mn^{2+} was incubated with rat brain tissue in the presence of catecholamine autooxidation products. It was found to inhibit cholinergic transmission, probably by being converted to a higher valency form under the effect of catecholamine autooxidation, causing LPO and receptor destruction and damaging the synaptic membrane. However, in rats Mn^{2+} increased the cholinergic transmission in the striatum in vivo owing to its scavenger action. LPO and ROIs affect dopaminergic transmission through their effect on the activity of the enzyme DBH, but also directly by acting on the synaptic membrane and on the receptors.

The inhibitory GABA-ergic neurotransmission may be stimulated with lipid-soluble antioxidants. In addition, ROIs may influence the modulation of neurotransmission. They may further inhibit the activity of Na^+K^+-ATPase, which indirectly controls neurotransmitter uptake, storage and release, inhibit acetylcholine esterase and partially also Ca^{2+}-ATPase. The phospholipase A_2 activating effect of ROIs and arachidonic acid released as a result also modulate synaptic membrane permeability.

LPO may be the neurochemical substrate of intelligence owing to its neurotransmission modulating effect. Intelligence is closely connected with short-term memory in which reversible, rapid changes in the permeability of the synaptic membrane, rich in PUFAs, play a significant role. This is indicated by the finding that lithium, characterized by a membrane depolarizing action, and ouabain, which inactivates Na^+, K^+-ATPase, inhibit short-term memory. IQ (intelligence quotient) showed a strong positive correlation with the glutathione peroxidase activity of red blood cells and a negative correlation with the cerebral lipofuscin level. A weaker positive correlation was found with the serum urate level (an increased frequency of exceptional intelligence had previously been observed among gout patients). The role of cholinergic transmission in memory has been proved. It has been known for some time that there is a positive correlation between learning ability of rats and mice and the acetylcholine esterase activity of their brains. Cholinergic transmission is defective in Alzheimer's disease: the cerebral levels of choline acetyltransferase and acetylcholine ersterase decrease. The decrease in the level of choline acetyltransferase is more specific because it is only found in the cholinergic neurones. Compared with acetylcholine esterase, it shows a better correlation with the severity of the disease and with the neuropathological changes. Acetylcholine esterase has been proved to inhibit the activity of ROIs [124, 482, 533, 575].

5.5 Psychiatric Syndromes

Increased red cell SOD activity has been found in a number of psychiatric syndromes like schizophrenia, manic depressive psychosis and Korsakoff's disease. If SOD activity is also higher in the brain, this may explain the decreased DBH activity described in schizophrenia, because the $O_2^-\cdot$ level needed for the enzyme's activity is decreased by the higher SOD activity. The increased activity of this enzyme may also play a role in the development of paranoia, hallucination, and Korsakoff's disease, by inhibiting the $O_2^-\cdot$-dependent enzyme indophenol dioxygenase, which leads to the inhibition of the decomposition of dimethyltryptamine, a hallucinogenic amino acid intermediate [1, 112, 404].

These propositions are supported by the observation that the mental handicap is less severe in Down's syndrome patients in whom the increase in SOD activity is moderate and, conversely, high SOD activity is associated with severe debility [404].

5.6 Cerebrovascular Syndromes and Trauma of the Central Nervous System

Free radical reactions probably also participate in the pathogenesis of regional ischaemia caused by cerebrovascular accident and acute trauma. In cats De-

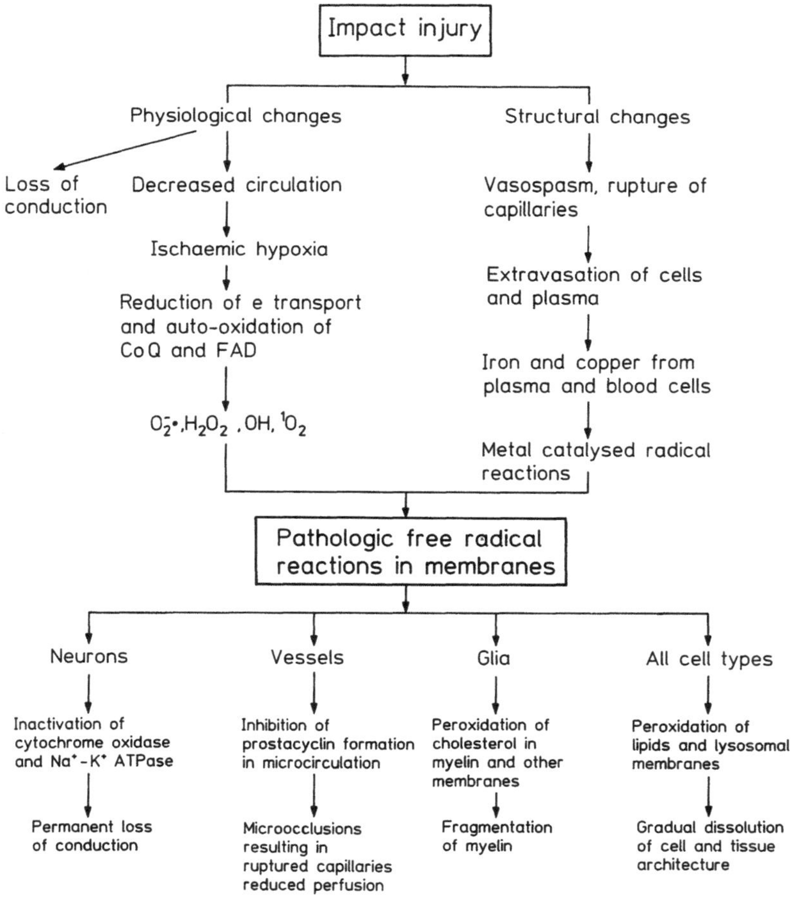

Fig. 28. Free radical pathomechanism of acute spinal trauma. CoQ coenzyme Q-ubiquinone; FAD flavine adenine dinucleotide. (After Demopoulos et al. [115])

mopoulos and co-workers ligated the medial cerebral artery unilaterally and thus produced the animal model of human stroke [115]. Similarly, the model of acute spinal chord trauma was obtained by dropping a weight on the spine of cats with intact dura. Although this is a rather far-fetched model, the changes observed correspond to those occurring in humans following accidents. Free radical reactions do not play a role in the early changes, which are reversible in both conditions (for approximately 3 h in the model), they are only involved in the late stage, when changes become irreversible. The mechanism of free radical formation in ischaemia is discussed in Sect. 8. In acute spinal trauma, part of the free radicals are formed as a result of ischaemia, while others are due to the extravasation of blood into the grey matter (Fig. 28). The concentration of PUFAs is highest in the grey matter, and iron (organic and inorganic) and copper get into the grey matter from blood, thus creating the conditions necessary for LPO. The selective inactivation of prostacyclin synthetase by hydroperoxides is the most important consequence which eventually leads to the irreversibility of the changes by causing microocclusions in the microcirculation [115].

The importance of ROIs in the pathogenesis of ischaemia has been confirmed by other authors [272, 568, 589]. Experimental ischaemic cerebral lesions were produced in animals by bilateral ligation of the carotid. Polyethylene glycol (PEG)- or Ficoll-conjugated SOD was effective when administered shortly before reperfusion and after it in association with allopurinol-inactivating xanthine oxidase. Injection into the brain of ROI-producing xanthine oxidase (hypoxanthine)-ADP-Fe^{3+} solution increased the permeability of the capillary endothelium and produced cellular (neurone, glia) damage and oedema. Using a similar ROI-producing system, it was shown that ROIs release arachidonic acid from synaptic membrane samples by activating phospholipase A_2 and cause membrane damage in which the ROIs produced during arachidonic acid metabolism may also play a role. Functional and morphological changes similar to those found in cerebral damage caused by acute hypertension and cerebral trauma could be produced in animal experiments by putting concentrated arachidonic acid on the brain surface. This also suggests that ROIs produced during arachidonic acid metabolism may play a role in all three types of cerebral damage [28, 74, 75, 224, 568].

In the majority of patients with cerebral trauma posttraumatic epilepsy develops. LPO may also play a role in this in accordance with the mechanism described in connection with Demopoulos' model of spinal trauma. These mechanisms were modelled by injecting $FeCl_2$ into the rat's cerebral cortex. An increased LPO was found in the brain. Although phenytoin stopped the epilectic seizures and EEG changes, it did not decrease LPO. Methylprednisolone, however, not only eliminated epileptic activity, but also decreased LPO significantly. In other experiments α-tocopherol and selenium pretreatment prevented epileptic activity and the increased LPO elicited by ferric salt injection. This suggests that after cerebral trauma, in addition to phenytoin, which eliminates the symptoms, glucocorticoids or other antioxidant substances acting on the hypothetical causative factor should also be administered [552, 583]. Lately several reports have been published on the favourable effect of large doses of corticosteroids and of barbiturates in central nervous system trauma and in cerebrovascular ischaemia. The antioxidant properties of these drugs may be at least partly responsible for this effect, moreover, the phospholipase A_2 inhibiting action of corticosteroids may also be involved [116, 506].

5.7 Cerebral Lipofuscinosis

Free radical reactions play a role in this rare neurological syndrome transmitted as a recessive autosomal trait. It is a fatal progressive disease of the central nervous system belonging to the group of amaurotic idiocies. Clinically the disease is characterized by mental handicap, seizures and disturbances of vision. These symptoms are probably connected with the accumulation of ceroid in the brain and retina. Pigment also accumulates in other organs. Inreased amounts of non-protein-bound iron and decreased antioxidant protection have been found in the cerebrospinal fluid of these patients. Ceroid accumulation is a sign of LPO, and CSF changes also indicate peroxidative damage. The deficiency of anti-LPO enzymes could not be unequivocally demonstrated, although some French authors found decreased linoleic acid and arachidonic acid levels (signs of peroxidative damage) associated with decreased lymphocytic SOD activity (this might account for the increased $O_2^{-}\cdot$ release and LPO). In infantile and juvenile forms of the disease decreased linoleic acid and docosahexanoic acid in the cerebral phospholipids have been found. The symptoms could be mitigated by the joint application of selenium and vitamin E [126, 217, 458, 472, 497].

5 Summary

The participation of free radical reactions has been found in several diseases of the CNS: in chronic organic brain syndrome, Alzheimer's disease, Parkinson's disease, in some psychiatric syndromes (schizophrenia, paranoid psychosis, Korsakoff's syndrome), in cerebrovascular diseases, trauma of the central nervous system, and in cerebral lipofuscinosis. The role of free radical reactions could not be clarified in any of these conditions because of the small number of experimental data. Most of these are connected with cerebrovascular diseases, traumatic central nervous damage and parkinsonism.

6 The Free Radical Pathogenesis of Atherosclerosis

Substances irritating the arterial wall increase atherosclerosis. The reactions of molecular oxygen with the PUFAs in serum and arterial wall are probably a constant source of such irritating substances [228, 229, 231, 232, 233]. In 1952, Glavind and co-workers described the presence of lipid peroxides in the human atherosclerotic aorta. At first they were believed to be artefacts of the process of preparation, but later this has been disproved [200, 273]. The peroxide content of the affected arteries was found to be proportional to the severity of atherosclerosis. LPO in the serum was demonstrated with ESR spectroscopy. The presence of ceroid in the intima of arteries might be regarded as a direct proof of LPO. The incidence of atherosclerosis increases with age. This may be due to the higher serum iron and copper levels, the increased amounts of unsaturated fatty acids in the lipids of the arterial wall, and their high degree of unsaturation found in the aged, which all favour LPO. Thus constant LPO in the serum, which increases with age, may initiate atherogenesis by damaging the arterial wall, and this predisposes to atherosclerotic lesions.

The important role of LPO in atherogenesis is shown by the epidemiological data. The average (fasting) serum copper level of white males who have suffered myocardial infarction is significantly increased. At the same time, normal men with higher serum copper concentrations are more susceptible to clinical atherosclerosis. Those consuming soft drinking water (with higher copper content) are more prone to atherosclerosis [228, 229, 231, 232, 233]. Hypercholesterolaemic patients are also likely to develop atherosclerosis, which is, at least partly, due to a shift in the direction of PUFA esters in the fatty acid esters of cholesterol, which is also normally present in blood in an esterified form. It is not cholesterol, but its oxidation products that initiate and then increase the damage of the endothelium of the arterial wall, which is one of the most common phenomena in atherosclerosis [229, 232, 271]. Muto and co-workers have shown in vitro that cholesterol is converted to its oxidation products under the effect of LPO [388]. In patients with various types of hyperlipidaemia the proportion of PUFAs was found to show an increase compared with the saturated fatty acids. An increased MDA level was found in both the plasma and the aortic wall [331]. Hyperlipidaemia may enhance atherosclerosis by other mechanisms, too. Lipoproteins are deposited in the vascular wall through its lesions. As has been shown (Chap. III Sect. 3), β-oxidation of fatty acids may also be found in the peroxisomes, in addition to the mitochondria. Increased lipoprotein deposition elicits compensatory peroxisome proliferation in the vessel wall, the aim of which is the decomposition of the lipids, H_2O_2 formed during the increased function of peroxisomes is insufficiently eliminated by the protective CAT mechanism, it leaves the peroxisomes, and may cause further damage of the vascular wall [87, 391].

LPO in the serum and in the vessel wall may also initiate the production of oxidation products of lipoproteins, depending on their PUFA content. This has been studied in the case of low-density lipoprotein (LDL) which has a high cholesterol content and is known to enhance atherosclerosis. The surface of the LDL molecule undergoes changes during its interaction with LPO products, which is demonstrated by its greater mobility during lipid electrophoresis. LDL which has undergone peroxidative modification is more readily phagocytosed by human monocytes and macrophages. This may explain the long-standing observation that diets with increased PUFA content decrease the serum cholesterol level. However, this should not be regarded as a favourable effect, because the decrease in the serum cholesterol level has its price: an increased deposition in the vessel wall of the cytotoxic, atherosclerosis-inducing oxidation products of LDL which have undergone excessive peroxidative changes. LDL incubated in vitro with smooth muscle cells of the vascular wall or with endothelial cells produced cytotoxic LPO products. Lipoproteins containing apolipoprotein B were investigated, and the presence in the arterial interstitium of toxic oxidative products having a higher electrophoretic mobility than plasma LDL was found. Summarizing these results, during the reaction of LDL with LPO products, cytotoxic substances are formed which may directly damage the vascular wall, and the macrophages of the vessel wall may increasingly phagocytose the modified LDL molecule. The amount of LDL incorporated by the macrophages may exceed the LDL degrading capacity of these cells and thus cytotoxic LDL products accumulate in the vessel wall [228, 248, 305, 378]. We found increased outflow of β-glucuronidase (lysosomal enzyme) from the granulocytes following cholesterol and triglyceride administration and in hyperlipidaemic patients, in whom the serum β-glucu-

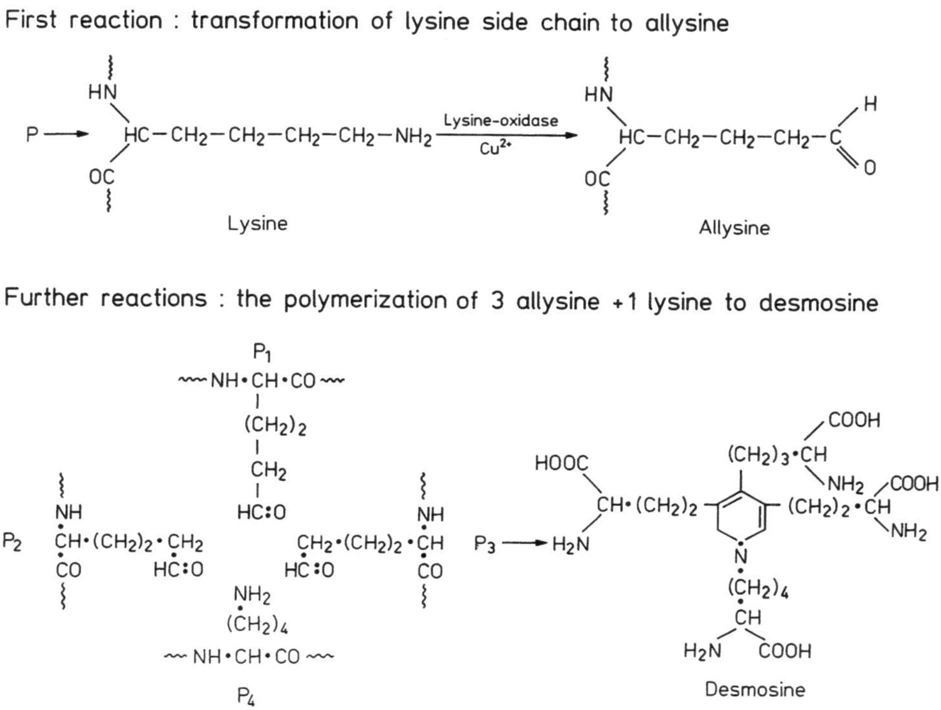

Fig. 29. Structure of the cross-links of elastin. (After Banga [20])

ronidase level also increased. These changes were regarded as the indirect signs of LPO [516, 561].

The famous Hungarian scientists J. Baló and I. Banga maintained that atherosclerosis in humans is associated with the destruction of elastic fibres. Banga demonstrated the presence of abnormal fluorescent lipofuscin-like and ceroid-like substances giving a positive reaction with TBA (indicating the presence of MDA) in intimal plaques and in the annulus fibrosus and nucleus pulposus of the intervertebral discs. The fluorescent substance could also be demonstrated in the serum of aged people in the lipoprotein fraction. It was TBA positive, especially in hyperlipidaemic cases. In the intimal plaque, accumulation of the fluorescent substance was found in the lipoprotein, which was coupled to elastin with strong bonds. Similarly to the results of Partridge [422], these data prove that there are lipoprotein molecules bound to elastin in atherosclerosis. These molecules bind calcium.

From these data, Banga proposed the following theory. Normally the polypeptide chains of elastin are held together by the cross-links (desmosine) between three allysine molecules and one lysine molecule (Fig. 29). The elastases of the aortic wall disrupt the cross-links, resulting in a reaction between MDA and other reactive aldehydes produced constantly during LPO in the serum on the one hand, and one or two lysine (or allysine) side chains on the other, be-

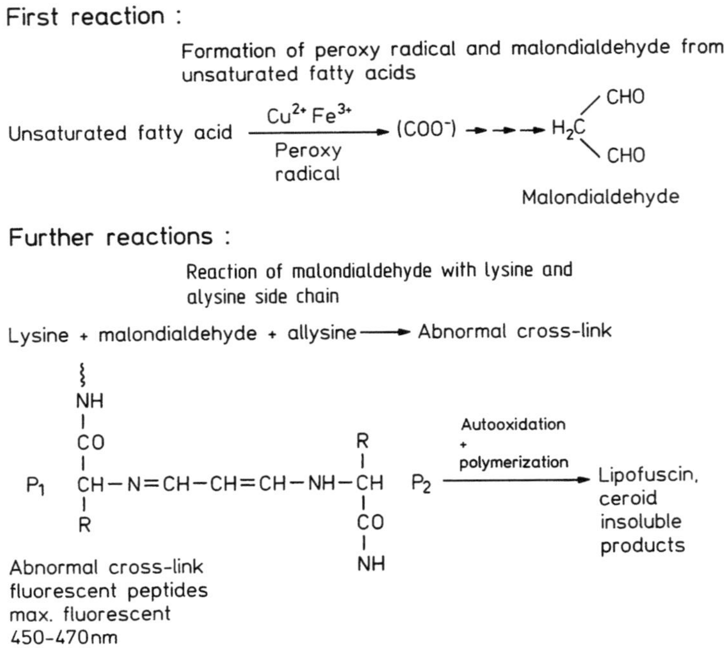

Fig. 30. Example of the cross-linking reaction of malondialdehyde with elastin. (After Banga [20])

cause the latter are not bound in desmosine. Their free state may result in abnormal cross-linkages which may contain several unsaturated bonds, leading to autooxidation and polymerization and the production of insoluble lipofuscin and ceroid products. Eventually a peroxidized lipid-calcium-elastin complex is produced (Fig. 30). Because of the abnormal cross-linkages, elastin loses its normal function, elasticity and water-binding capacity, and calcium is deposited in the vessel wall, which becomes hardened [20].

Since this theory was described important advances have been made in atherosclerosis research: the role of the PGI_2-TXA_2 balance has been recognized. By activating phospholipase A_2, pathological free radical reactions stimulate arachidonic acid metabolism. This would lead to increased synthesis of both TXA_2 and PGI_2, but it has been proved that lipid peroxides can selectively inhibit PGI_2 synthesis in the vessel wall. Thus pathological free radical reactions increase TXA_2 synthesis, thereby causing a shift in the PGI_2/TXA_2 balance in favour of TXA_2. As a result, the aggregation of platelets and their adhesion to the vessel wall increase (Fig. 31).

Indeed, in the aorta and heart of rabbits made atherosclerotic, decreased amounts of PGI_2 were produced. Similar results were obtained in human atherosclerotic tissue in which no PGI_2 production could be demonstrated, while in neighbouring normal vessels PGI_2 synthesis was found. PGI_2 synthetase is probably inactivated by $\cdot OH$ formed during the interaction of lipid peroxides.

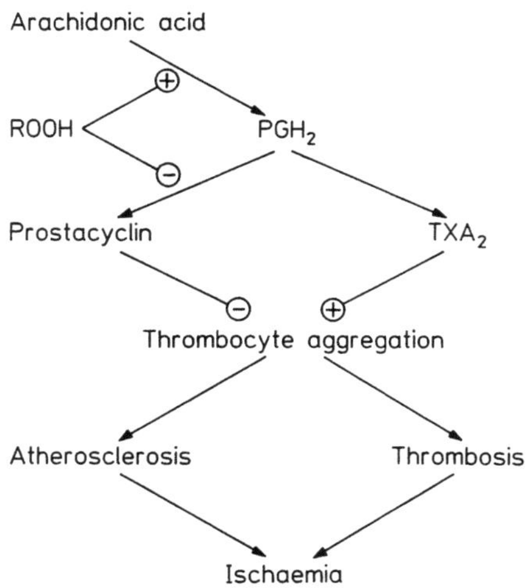

Fig. 31. Effect of lipid hydroperoxides on arachidonic acid metabolism and the pathophysiological consequences. PGH_2 prostaglandin H_2; TXA_2 thromboxane A_2. (After Warso and Lands [566])

Increased adhesion of platelets to the vessel wall results in their increased reactivity; platelet-produced substances (growth factor, $O_2^{-}\cdot$) probably play a role in the ensuing damage of the vessel wall. Thus it is understandable that even the early atherosclerotic lesions (fatty streaks) lead to a decrease of PGI_2 synthesis in the vessel wall [216, 222, 373, 374, 566, 574].

The protease inhibitor role of free radical reactions has been proved (see Sect. 14). They also inhibit the activity of the serum elastase inhibitor, as a result of which the first step described in Banga's theory, i.e. the activation of tissue elastases, may occur. The role of lipid peroxides was substantiated by the experiment in which a vitamin E-deficient diet caused a decrease in PGI_2 content of isolated segments of the aorta of rabbits. The decrease could be reversed by vitamin E administration [73]. Smoking has also been shown to increase platelet aggregation, which is probably due to LPO induction. This may, at least in part, be responsible for the vascular complications caused by smoking [80]. The free radical theory of atherosclerosis is especially attractive because it reconciles two seemingly contradictory theories, i.e. the lipid theory[1] and the thrombogenic theory[2] [374] (Fig. 32).

In severe atherosclerosis granulocytes probably also contribute to the damage of the vessel wall. Ulcerating plaques allow crystalline cholesterol,[3] which normally does not get into contact with plasma, to activate the comple-

---

[1] This theory attributes primary importance to disturbances in the lipid metabolism as a result of which lipids are deposited in the intima of the vessels.

[2] This theory maintains that the atherosclerotic plaques are organized microthrombi originating from platelets arrested by microtraumas of the intima. The vasoactive substances released by the trapped platelets increase the permeability of the vessel wall locally, leading to lipid infiltration.

[3] Crystalline cholesterol is formed if the three-phase water-phospholipid-cholesterol system becomes saturated with cholesterol, and the precipitation of cholesterol molecules leads to the formation of cholesterol monohydrate crystals.

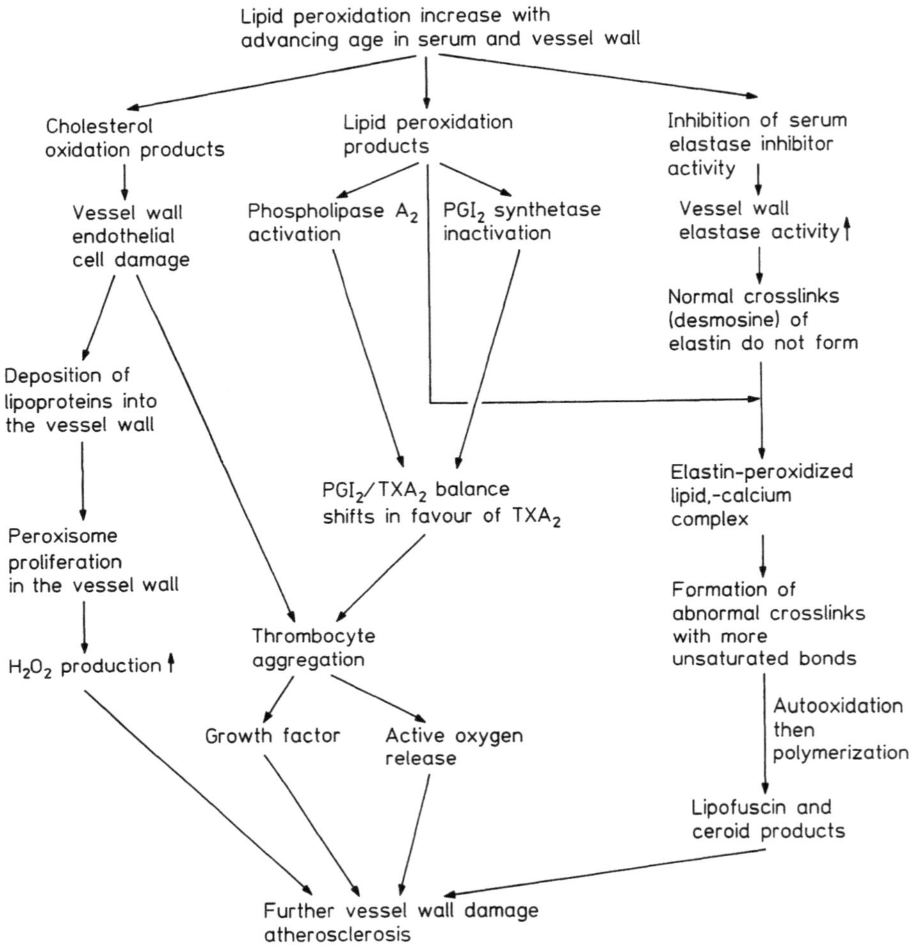

Fig. 32. Free radical reactions in the pathogenesis of atherosclerosis, for explanation see text

ment system. C5a causes aggregation of granulocytes and increases the adhesion of granulocytes to the vessel wall and the damage of the vessel wall by activating the respiratory burst. A similar mechanism is probably activated if there is haemorrhage into the plaques, another occasion when crystalline cholesterol gets into contact with plasma. This mechanism not only increases the severity of the local process, but also causes leukostasis and the formation of leukothrombi leading to ischaemia in the small vessels which are in the direction of the flow and are still unobstructed [223].

It has been attempted to use antioxidants in the therapy of atherosclerosis. Vitamin E administration in angina pectoris produced varying results, which was attributed to variations among the studies in the number of patients and the duration of treatment [10]. Vitamin E was used with success in the treatment of intermittent claudication, it improved both the subjective and objective symptoms [218, 219]. We, too, found that antioxidants have a favourable effect in atherosclerosis. MTDQ-DS (for structure and effect see Chap. VII Sect. 2), a water-soluble dihydroquinoline-type antioxidant, was effective in atherosclerosis produced with cholesterol in rabbits. We also have preliminary data indicating that this agent has a favourable effect in human hyperlipoproteinaemia and obliterative atherosclerosis. MTDQ-DS significantly decreased the serum cholesterol, LDL, very low density lipoprotein (VLDL) and triglyceride levels and increased the high-density lipoprotein (HDL) fractions in animal experiments and in a patient with type IV asymptomatic hyperlipoproteinaemia. The atherosclerosis index (LDL/HDL or LDL + VDL) also decreased. In animal experiments treatment with cholesterol + MTDQ-DS decreased the rate of development of aortic plaques by 90% indicating the inhibitory effect of the drug on cholesterol degradation and this endothelium protective effect.

In a patient suffering from symptomatic obliterative atherosclerosis complicated by thrombosis of the popliteal artery, in whom surgery was not necessary because collateral circulation was maintained, MTDQ-DS administration improved the results of traditional treatment (vasodilator, anticoagulant) in respect of the objective signs and walking distance. The treatment influenced the serum lipid levels and the atherosclerosis index in the same way as in the patient with asymptomatic hyperlipoproteinaemia. We found in animal experiments that MTDQ-DS decreases the elevated serum and platelet MDA levels, which also indicates that it decreases the oxidative degradation of cholesterol and inhibits the platelet TXA_2 synthesis. Thus MTDQ-DS, and antioxidants in general, may be used as new specific drugs in the treatment of atherosclerosis [152, 155, 156, 198, 606]. The effect of antioxidant (vitamin E) therapy on blood lipids has been described by several authors [23, 250, 527]. The dihydroquinoline-type antioxidant was also suitable for preventing the damage caused by cholesterol in the membrane of hepatic lysosomes [546]. Prostacyclin therapy is another possible approach. Its favourable effect is the subject of several reports [326, 373, 374].

Dietary measures may be used with success. The incidence of acute myocardial infarction is lower in Eskimos, which is attributed to their being less prone to atherogenesis. Low levels of serum cholesterol, triglyceride, LDL and VLDL were found in their serum, which in men was associated with a high HDL level. At the same time their haemorrhagic tendency was increased. These changes were attributed to eicosapentaenic acid (20: 5w3, EPA), a component of the Eskimos' seafood diet. During its metabolism TXA_3 is produced in the platelets, which, unlike the TXA_2 derived from arachidonic acid, has no platelet aggregating effect. At the same time, PGI_3 is produced in the vessel wall. It has an antiaggregation effect similar to that of PGI_2 originating from arachidonic acid. Thus a high level of EPA associated with a low arachidonic acid level results in an antithrombotic state and is also responsible for the increased haemorrhagic tendency. In this light the earlier dogma, according to which the antiatherosclerotic diet should contain little saturated lipid and much unsaturated lipid, must be revised. The ideal diet has a high EPA content, but large amounts of unsaturated lipids exceeding the body's essential fatty acid requirement are harmful and should be avoided because they increase LPO [135, 138, 265).

6.1 Summary

Molecular oxygen reacts with the PUFAs in the serum and aortic wall, leading to LPO. With advancing age this tendency increases, because the serum iron and copper levels increase and there is an increase in the amount and degree of saturation of PUFAs in the lipids of the arterial wall. The atherogenic effect of hyperlipoproteinaemia is due to an increase in the proportion of PUFAs prone to peroxidation. Under the effect of LPO in the serum and vessel walls, these PUFAs are more easily converted to oxidation products which damage the vascular wall. Through the lesions lipoproteins are deposited in the vessel wall and elicit compensatory proliferation of peroxisomes. H_2O_2 *produced during increased peroxisomal activity, when exceeding the protective potential of CAT, leaves the peroxisomes and may cause further damage of the vessel wall.*

According to the free radical theory of atherosclerosis, free radical reactions inhibit the serum elastase inhibitor and prostacyclin synthetase activity, and by activating phospholipase A_2, *enhance* TXA_2 *formation. Because of the inhibition of the serum elastase inhibitor, the tissue elastases of the aortic wall are activated, the cross-links of elastin are broken, and the lysine and allysine side chains thus released interact with the reactive products of LPO in the serum and aortic wall. Then they undergo autooxidation and polymerization, producing insoluble lipofuscin and ceroid products, which correspond to a peroxidized lipid-calcium-elastin complex. Prostacyclin synthesis decreases or is absent because the activity of prostacyclin synthetase is inhibited. This and the increased* TXA_2 *production, which is due to the activation of phospholipase* A_2, *increase the adhesion of platelets to the vessel wall, and the ROIs produced further increase the damage. The free radical theory reconciles the earlier, contradictory, lipid and thrombogenic theories. Antioxidants have been used with success in the treatment of atherosclerosis in animal experiments and in humans. The authors' own results also prove this effect. Pathological free radical reactions thus play a causative role in the development of atherosclerosis, as shown by experimental evidence. The free radical theory is currently one of the most modern theories trying to explain the pathogenesis of atherosclerosis.*

7 Hypertension and Free Radical Reactions

The increase of peripheral resistance in the small blood vessels, especially arterioles, is the most important factor in hypertension. The irritating effect of peroxidative reactions (occasionally by releasing various PG derivatives) causes contraction of the blood vessels. Progressive arteriocapillary fibrosis, developing due to fibroblast stimulation following vascular irritation (see Sect. 3) causes fixed, progressively higher blood pressure. Inhibition by lipid peroxides of the synthesis of prostacyclin, which decreases vascular tone, may also contribute to the development of hypertension. Essential hypertension may, at least partially, be related to factors increasing atherosclerosis with the advance of age, i.e. with LPO.

Dilatation of the cerebral arterioles, damage of the arterial endothelium and of the blood-brain barrier have been found in acute hypertension in experimental animals and in cerebral trauma. In both cases the functional and morphological changes could be prevented by pretreatment with indomethacin and AHR-5850 inhibiting cyclooxygenase. By applying arachidonic acid or the ROI-producing xanthine oxidase-xanthine system to the cerebral surface, lesions closely resembling the changes caused by acute hypertension and cerebral trauma could be produced. The same effect was achieved with PGG_2 and 15-hydroperoxyeicosatetraenic acid (15-HPETE) application, whereas PGH_2 did not produce vascular damage, only vasodilation. The changes caused by acute hypertension and cerebral trauma could be prevented by pretreatment with scavengers (SOD, nitro blue tetrazolium, mannitol), while the damaging effect of arachidonic acid and the xanthine oxidase-xanthine system could be prevented by SOD and/or CAT administration. These data suggest that arachidonic acid metabolism and the ROIs produced during it - or originating from other sources - may play a role in the pathogenesis of acute hypertension and cerebral trauma (for the latter see also Sect. 5.6). Arachidonic acid metabolism is stimulated by activation of phospholipase A_2. The mechanism whereby this occurs is not known, but free radicals may also serve as the trigger factor. Hence antioxidants may have a favourable effect in hypertensive encephalopathy.

Peroxidative processes probably also play a role in the pathogenesis of toxaemia in pregnancy. The serum copper and lipid levels increase in pregnancy, while that of ascorbic acid decreases. Products of free radical reactions have been found to accumulate during pregnancy; other authors report on the accumulation of the conjugated diene isomer of linoleic acid. These changes occurring in normal pregnancy are significantly escalated in toxaemia. In addition, a decrease in the prostacyclin levels was found in the wall of the maternal and fetal blood vessels, which was due to the inhibitory effect of lipid peroxides. A correlation could be demonstrated in preeclampsia between the elevated serum lipid levels and the increase in blood pressure. Significant elevation of the level of LPO products as compared with normal pregnancy may make this test suitable for predicting preeclampsia [143, 214, 228, 229, 232, 233].

7.1 Summary

The atherosclerosis-inducing effect of free radical reactions, their relation with arachidonic acid metabolism, and their inhibitory effect on prostacyclin, which decreases vascular tone, may play a role in essential hypertension. However, their part in the pathogenesis of hypertension needs further clarification, and the relevant experimental data are insufficient for drawing conclusions. Preliminary data suggest the involvement of pathological free radical reactions in the pathogenesis of toxaemia in pregnancy.

8 Free Radical Reactions in Ischaemic Damage

If tissue blood flow decreases or ceases, the immediate result is an increase in the permeability of the capillaries leading to oedema. In case of severe injury, microscopic or gross tissue destruction may be found. This damage occurs not so much in the ischaemic phase as during reperfusion. Granger and co-workers [210] were the first to prove the role of free radical reactions in an intestinal ischaemia model in the cat. They measured capillary permeability and found that intravenous SOD pretreatment, unlike diphenhydramine and cimetidine, indomethacin or methylprednisolone, significanly reduced the increase in capillary permeability caused by ischaemia. In another experiment they increased the time of ischaemia and studied the effect of intravenous SOD and oral allopurinol pretreatment on the microscopic mucosal changes. They found a significant improvement compared with the controls. They used allopurinol because, on the basis of a previous experiment they postulated that xanthine oxidase may be the source of the $O_2^-\cdot$ produced in the ischaemic intestines. This idea had been supported by the protective effect of allopurinol described in haemorrhagic shock and myocardial ischaemia. Xanthine oxidase was the first documented biological source of $O_2^-\cdot$. The intestines, lungs and liver contain especially high levels of the enzyme, although it has been proved that in vivo the enzyme is not an $O_2^-\cdot$-producing oxidase (O type)

$$\text{Hypoxanthine} + H_2O_2 + 2\,O_2 \xrightarrow{\text{O type}} \text{Xanthine} + 2\,O_2^-\cdot + 2\,H^+$$
$$\text{Xanthine} + H_2O_2 + 2\,O_2 \xrightarrow{\text{O type}} \text{Urate} + 2\,O_2^-\cdot + 2\,H^+$$

but rather an NAD^+-reducing dehydrogenase (D type)

$$\text{Hypoxanthine} + H_2O_2 + NAD^+ \xrightarrow{\text{D type}} \text{Xanthine} + NADH + H^+$$
$$\text{Xanthine} + H_2O_2 + NAD^+ \xrightarrow{\text{D type}} \text{Urate} + NADH + H^+$$

If the tissues are homogenized, the D type is converted to the O type either reversibly via sulphydryl oxidation, or irreversibly via limited proteolysis. But the O type is not only a preparative artefact. It has been shown that ischaemia causes a rapid D→O conversion in situ. In rats this occurred irreversibly, i.e. by proteolyis.

Since the energy content of hypoxic cells decreases rapidly and the maintenance of the cation gradient is an energy-dependent process, hypoxia probably leads to calcium influx, and this activates the calmodulin-regulated proteolytic enzymes. This is indicated by the finding that trifluoperazine, which has an inhibitory effect on calmodulin, strongly decreased the D→O conversion, as well as by the phenomenon of calcium paradox (see later discussion). The appearance of xanthine oxidase in ischaemic tissue is not sufficient for the development of free radical damage, because substrates are also needed. The accumulation of the first substrate, hypoxanthine, as a result of ATP catabolism has been demonstrated in ischaemic tissues. In hypoxia ATP is dephosphorylated and the cellular AMP strongly increases. Then AMP is catabolized to adenosine, inosine and finally to hypoxanthine. For O-type activity another substrate, O_2, is needed, most of which enters the tissue during reperfusion and gives rise to $O_2^{-}\cdot$ formation. This explains why the more severe lesions occur only after reperfusion [42, 54, 104, 210, 359, 420] (Fig. 33).

Dalsing and co-workers proved this hypothesis from another aspect in model experiments with rats with intestinal ischaemia. Administration of aminophylline (a xanthine derivative), which may be the substrate of xanthine oxidase and thus induce $O_2^{-}\cdot$ formation, increased the animals' mortality significantly and enhanced the morphological changes of ischaemia. Both mortality and morphological changes decreased significantly if SOD was administered at the same time [105]. The hypothesis also explains why soybean trypsin inhibitor, which inhibits proteolysis, decreased the ischaemic mucosal lesions elicited in dogs by occlusion of the splanchnic artery and prevented the increase in vascular permeability caused by intestinal ischaemia in cats. In the same way, it accounts for the inhibitory effect of the calcium antagonist verapamil on damage of the cardiovascular system and small intestinal mucosa in dogs exposed to haemorrhagic shock [420]. Parks and co-workers have demonstrated, similarly to Del Maestro and other authors [111, 327], that ·OH formed secondarily from $O_2^{-}\cdot$ via the Haber-Weiss reaction may be responsible for the intestinal vascular permeability changes found in ischaemia, because pretreatment with the ·OH scavenger dimethylsulphoxide (DMSO) provided a protection comparable to that observed in cases of SOD pretreatment [421].

The ischaemic lesions of the large intestine cannot be attributed to the xanthine oxidase mechanism, because, unlike in the small intestine, xanthine oxidase activity is relatively low in the large intestine. However, the large intestinal mucosa is rich in aldehyde oxidase, which can release ROIs. This may be the alternative free radical mechanism of ischaemic damage of the large intestine [420].

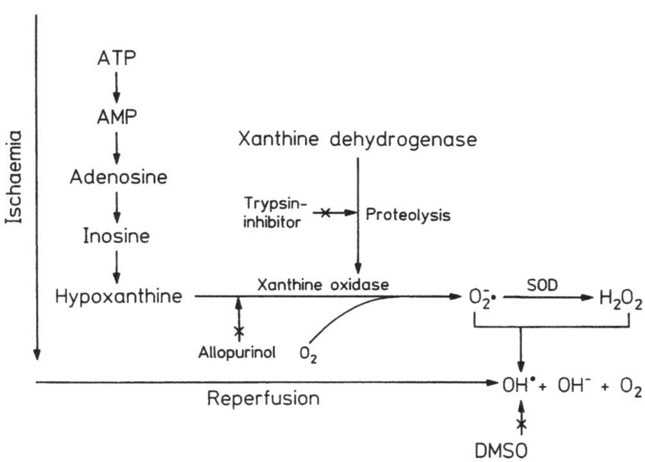

Fig. 33. Free radical mechanism of ischaemic damage, for explanation see text. DMSO dimethylsulphoxide; *crossed arrow* inhibition. (After Parks et al. [420])

Based on the model, the varying sensitivity of different tissues to ischaemia may be attributed to differences in their xanthine dehydrogenase content, in their level of protease, which induces the D→O conversion, and in the function of the individual proteases. In rats conversion to oxidase was quickest in the ileum, followed by the heart, liver, spleen, lungs and kidneys. Xanthine dehydrogenase was also demonstrated in skeletal muscles; this, however, was not converted to oxidase when perfusion was discontinued (probably owing to the altered function, decreased amount or absence of protease). This observation correlates with the known relative resistance of skeletal muscles to ischaemic injury in comparison with other tissues. Tissue-dependent differences have also been found in other respects: free radical scavengers and allopurinol can protect the small intestine only against partial ischaemia, whereas in the heart and skin they are effective against global ischaemia, if this is kept within limits. This shows that in global ischaemia other mechanisms independent of the free radical reactions are dominant in the small intestines, while in the heart and skin global ischaemia differs from partial ischaemia only quantitatively up to a limit, which depends on the capacity of antioxidant protection. This phenomenon has not been fully explained [358].

The free radical mechanism lends itself to the explanation of not only the paradoxical deterioration of the ischaemic lesions after reperfusion (oxygen paradox), but also to that of the histologically similar phenomenon of calcium paradox occurring if the myocardium is first perfused with a calcium-free solution and then with a solution containing calcium. Since the histological picture is similar in both cases and both processes have two similar steps (lesions caused by the deprivation of a substance are paradoxically aggravated by the reintroduction of this substance), it has been suggested that these phenomena represent two facets of the same problem. This has been supported by the finding that the perfusion of rat hearts with cumene hydroperoxide (a synthetic lipid hydroperoxide), which causes LPO in the heart, produced changes that were similar to those elicited by the calcium paradox, namely cellular protein release, decreased coronary flow, and myocardial contraction. The ·OH scavenger DMSO protected against the calcium paradox injury. An elevated MDA level was found in the rat heart in coronary perfusate in the course of calcium paradox injury. The exact mechanism of the calcium paradox is not known. One of the possibilities is the activation of membrane phospholipases on reintroduction of calcium. This triggers the arachidonic acid metabolic cascade, during which LPO-inducing ROIs are released. According to another hypothesis, reintroduction of calcium activates calmodulin, which accelerates the D→O conversion and thus also ROI production. It also activates the membrane phospholipases, which trigger the sequence of reactions discussed previously [255, 288].

These reactions play a part, in addition to ischaemic intestinal diseases, in haemorrhagic and endotoxin shock (perhaps also in other types of shock), in myocardial ischaemia, in cerebrovascular events associated with ischaemia, and in the pathogenesis of reperfusion damage following organ transplantation. Finally, $O_2^-\cdot$ may also be formed owing to hypoxia in the respiratory chain via another mechanism (see Chap. III Sect. 3).

8.1 Ischaemic Cerebrovascular Disease

Ischaemic cerebrovascular disease is discussed in Sect. 5.5.

8.2 Myocardial Ischaemia and Infarction

There is abundant evidence available proving the role of free radicals in myocardial ischaemia. Japanese authors have demonstrated the increase of LPO in patients with myocardial infarction; they found an increase in TBA-positive substances in the blood of these patients [298]. Soviet authors used a more direct method (chemiluminescence) in their model experiments in rats, and found that the degree of myocardial damage was proportional to the intensity of LPO [309]. It has long been known that in global ischaemia lasting for a short time (e.g. during cardiac surgery) and in the early stage of myocardial infarction, lipofuscin accumulates in the myocardium [309]. There are data available on the protective effect of scavengers and antioxidants in myocardial infarction. In rats sodium selenite, vitamin E and also their combined use decreased the elevated LPO, improved the ECG changes and decreased myocardial damage [310]. Free radical scavengers like SOD, CAT and mannitol (·OH scavenger) and antioxidants allowed a better restoration of left ventricular function after reperfusion [191, 252, 494]. Casale and co-workers found that the protective effect of scavengers is not weaker if they are given during early reperfusion as compared with administration at the beginning of ischaemia, indicating that the damage occurs predominantly during early reperfusion [191]. Structural evidence for this has been provided by Reimer and co-workers and by Jennings and co-workers [282, 452]. Others added systems producing ROIs directly (xanthine oxidase-purine) to perfused intraventricular septum of rabbits and demonstrated free radical induced myocardial damage [56]. The incidence of arrhythmias during reperfusion could be reduced with scavengers and allopurinol [244, 338].

The results of the global ischaemia model experiments have been utilized in cardiac surgery: damage caused by reperfusion after the complete ischaemia necessary during surgical intervention could be reduced. Chambers and co-workers produced myocardial infarction in animals by transient ligation of the descending ramus of the left coronary artery and found that both SOD and allopurinol (the former more strongly) decreased the infarcted area [71, 72]. Similar results were described by other authors [244, 262, 287, 336]. Based on these data, administration of scavengers and/or allopurinol before or during reperfusion may be recommended for decreasing reperfusion damage and arrhythmia in systemic and intracoronary thrombolysis and percutaneous transluminal angioplasty [72, 173, 244, 262, 287, 336, 338].

The role of LPO in the pathogenesis of myocardial ischaemia and infarction has been described and proved experimentally by Hungarian authors [471, 550].

Hess and co-workers attribute a scavenger effect to the glucose-insulin-potassium (GIK) infusion applied by Sodi-Pallares during cardiac surgery in normothermic regional ischaemia and in global ischaemia and hypothermia

[254]. It had been shown earlier that in both normothermic and hypothermic ischaemia as well as during reperfusion following hypothermic ischaemia a defect of the excitation and contractile apparatus of the myocardium may be found. The reason for this is supposed to lie in the uncoupling of the calcium transport of the sarcoplasmic reticulum and ATP hydrolysis, which has been attributed to the free radical induced damage. The role of ·OH is believed to be prominent in this damage. The protective effect of SOD and mannitol points in the same direction. Uncoupling is also inhibited by the GIK infusion. Glucose, similarly to mannitol, is probably an ·OH scavenger, while insulin promotes its entry into the cells [252, 254].

In the early phase of myocardial infarction neutrophilic granulocytes migrate into the injured tissue and their activation is associated with ROI release in addition to lysosomal enzyme release. Thus free radicals can indirectly increase the severity of myocardial damage. The chemotactic factor (for the mechanism of its production see Sect. 2) stimulates the migration of granulocytes. Thus it is not surprising that in myocardial infarction model experiments dogs made neutropenic with neutrophilic antiserum had an infarcted area which was 43% smaller than that found in animals with normal neutrophil counts [254, 275, 354, 465]. SOD also inhibits this indirect damaging effect (cf. Sect. 2). In another experiment neutrophils activated with phorbol myristate acetate (PMA) damaged cardiac function significantly (decreased the cardiac output and arterial pressure and increased peripheral resistance and left ventricular and diastolic pressure). SOD and CAT administration and neutropenia produced by antineutrophilic antibodies had a protective effect [469].

8.3 Renal ischaemia

A protective effect of free radical scavengers (SOD, CAT, mannitol and the ·OH scavenger dimethylthiourea) and allopurinol was demonstrated in post-ischaemic reperfusion renal damage [226, 409, 415, 483]. The degree of tubular cellular damage decreased in response to mannitol and chlorpromazine (·OH scavengers) during reperfusion of ischaemic rabbit kidneys [419]. The protective effect of mannitol, which has been used in the treatment of acute renal failure for a long time, is probably due to its ·OH scavenger activity, in addition to its increasing effect on renal blood flow and solute diuresis [337].

8.4 Shock

In shock, selective vasoconstriction occurs, which affects the blood vessels of abdominal organs in the first place, while the blood supply of vital organs (brain and heart) is unaffected. Owing to hypoxia, ATP is decomposed in the abdominal organs and the hypoxanthine produced gets into the plasma. Free radical reactions may have a part in the shock becoming irreversible, because a late restoration of the intravascular volume only adds to the severity of the shock and causes reperfusion damage. Disturbed microcirculation found in irreversible shock, i.e. increased capillary permeability, endothelial damage and oedema, resemble the vascular damage produced by free radical induced ischaemia. Allopurinol and verapamil were effective in the haemorrhagic shock of dogs, while cysteine, which also has ·OH scavenger action, is effective in cardiogenic shock [419]. In experimental endotoxin shock induced with a single endotoxin injection the serum levels of lysosomal enzymes, believed to be sensitive indicators of free radical induced damage and of TBA-reactive products, increased. SOD pretreatment prevented the development of shock. SOD and CAT had a protective effect in experimental disseminated intravascular coagulation induced with prolonged endotoxin infusion, and decreased the level of TBA-reactive products in the serum significantly [592]. Phenyl-*t*-butylnitrone (PBN), a scavenger used in spin trapping studies, provided effective protection in experimental traumatic and endotoxin-induced shock and in shock induced with the ligation of the superior mesenteric artery [406].

8.5 Transplantation

The ischaemia occurring during the preservation of organs for transplantation, during transplantation itself, and during the subsequent reperfusion resembles the conditions encountered in experiments in which free radical reactions are elicited. For example, a favourable effect is achieved by adding antioxidant steroids, chlorpromazine and allopurinol to the solution in which cadaver kidneys are stored until transplantation. Free radical reactions induced by reperfusion damage may play a role in acute tubular necrosis that occasionally follows kidney transplantation. It has been proved experimentally that allopurinol pretreatment improves posttransplantational renal function and graft survival [419].

SOD or allopurinol have been found to prevent the damaging effect of complete anoxia occurring if the supplying artery of cutaneous island flaps is occluded. The pedicle of island flaps used in plastic surgery contains only the direct supplying artery of the skin flap, hence there is no collateral circulation. Application of antioxidants may promote the more extensive use of free (unpedicled) flaps in plastic surgery [419].

8.6 Summary

Free radical reactions probably play a causal role in the pathogenesis of ischaemia-induced damage. Paradoxically, the damage occurs not in the ischaemic phase, but during reperfusion. Pathological free radical reactions also occur predominantly in this phase. ROIs are produced by the mediation of cytoplasmic xanthine oxidase, which, in fact, is an NAD^+*-reducing dehydrogenase under physiological conditions and not an oxidase. Under the effect of ischaemia, however, it is converted to oxidase, producing* $O_2^-\cdot$ *from hypoxanthine or xanthine substrate and oxygen. Hypoxanthine is a product of the ATP catabolism of ischaemic tissues. Oxygen, the other substrate, is available at reperfusion. This hypothesis has been experimentally supported. The mechanism described may have a role in the following diseases: intestinal ischaemia, myocardial infarction, shock, cerebrovascular events associated with* *ischaemia, renal ischaemia, preservation of organs for transplantation, reperfusion damage following organ transplantation. A favourable effect of antioxidants and allopurinol, which inhibits xanthine oxidase, was observed in each of these conditions. It is in the antioxidant and allopurinol therapy of myocardial infarction and intestinal ischaemia that we have the highest number of data.*

9 Free Radical Reactions in Other, Nonischaemic Heart Diseases

Large doses of catecholamines are known to cause myocardial necrosis. Catecholamines readily undergo autooxidation, which leads to ROI production. The ROIs and the oxidation products of catecholamines (e.g. adrenochrome and other oxidized catecholamine metabolites), rather than catecholamine itself, are responsible for the damage. Myocardial damage is, at least in part, due to the LPO elicited by the ROIs (see also the adrenergic neuronal damage caused by 6-OHDA, Sect. 5) [449, 499, 532]. Myocardial cell damage, ECG changes and high mortality induced in rats with isoproterenol could be prevented by the administration of vitamin E and zinc, which has a membrane-stabilizing effect [499].

Ramos and co-workers treated rat myocytes in culture with toxic doses of isoproterenol and found that at first the mitochondria were damaged, then the sarcolemma, the latter inducing an increase in intracellular potassium and LDH release. The increase in intracellular potassium is due to the inhibition of Na^+,K^+-ATPase. The toxic effect of isoproterenol can be counteracted by the administration of antioxidants (ascorbic acid, sodium bisulphite) [2, 449]. The increasing effect on myocardial LPO of a strong increase of catecholamine in severe emotional or physical stress has also been described [61]. Capel and co-workers exposed rats to cold stress and found a significantly increased lipid peroxide level in the animals' liver and plasma. The hepatic GP and glutathione reductase levels also increased. They attributed the increased LPO to the stimulation of the microsomal mixed-function oxidase system as a result of stress-induced hormonal changes [61]. Free radical reactions thus play a role in the pathogenesis of stress, and indirectly in the development of myocardial ischaemia and infarction, early atherosclerosis and stress ulcer. In ischaemic heart disease there is an increased outflow of catecholamine, therefore, the damaging effect of catecholamines also plays a role in the pathogenesis.

The combined deficiency of selenium and vitamin E causes degenerative heart disease in various species of domestic animals. A heart disease characterized by focal myocardial necrosis without coronary damage, which was lethal in a high percentage of the cases, was described in China's Keshan province. Improvement in this disease could be achieved with selenium therapy.

The serum selenium and SeGP concentration of the patients was very low. A prophylactic effect of selenium was also observed. Clinically the disease corresponds to congestive cardiomyopathy (COCM), but the aetiology is still unclear and there are epidemiological characteristics which cannot be attributed to selenium deficiency alone. The role of Coxsackie B4 virus has also been suggested. Selenium-deficient mice were found to be more sensitive to the damage caused by this virus [76, 322, 596]. Evidently, the role of selenium in idiopathic COCM has been suggested. Low selenium levels in two cases of COCM were described in western countries. In one of these, long-term parenteral feeding, involving selenium intake, was at least partially responsible for the development of COCM. The patient was given selenium for 3 weeks during which time it was found ineffective. Eventually the patient died amid recurrent attacks of arrhythmia. In the other case 4 weeks selenium therapy improved the symptoms, but cardiomegaly persisted even after 3 months [85, 285].

Other factors supposed to play a role in the aetiology of COCM, e.g. alcohol and pregnancy, increase LPO, similarly to selenium deficiency. Therefore, the role of free radical reactions in the disease might arise, although it has not been proved. The light- and electron-microscopic picture of alcoholic myopathy and cardiomyopathy resembles that of myopathy and cardiomyopathy of LPO origin in vitamin E-deficient domestic animals. In both cases the mitochondria, rich in PUFAs, and the sarcoplasmic reticulum are primarily damaged. The toxic effect of alcohol observed after several years of regular consumption is supposed to be due to the alcohol metabolite acetaldehyde. It may cause LPO damage via two mechanisms: (a) $O_2^{-}\cdot$ is formed during the xanthine oxidase-metabolized oxidation of acetaldehyde, or (b) acetaldehyde is bound to the SH-containing precursors of GSH and thereby decreases the GSH concentration (this is supported by the protective effect of the GSH precursor L-cysteine against the toxic effects of acetaldehyde) [189].

9.1 Summary

ROIs are released during the autooxidation of catecholamines. ROIs and oxidized catecholamine metabolites are supposed to be responsible for experimental myocardial necrosis induced with catecholamines. Increased ROI production is induced by hormonal changes stimulating the microsomal mixed-function oxidase system in stress, and by catecholamines accumulating in stress. Thus ROIs may play a role in the pathogenesis of stress. Their role has been proved in the Keshan disease described in China, which is a frequently lethal necrosis of the myocardium. Clinically it resembles congestive cardiomyopathy. Selenium deficiency and lower selenium-dependent GP activity was found in these patients. The disease may be prevented by selenium prophylaxis. A role for free radical reactions in congestive cardiomyopathy has been suggested, but not yet proved, since alcohol, which causes pathological free radical reactions, is involved in the aetiology of the disease.

10. The Role of Free Radical Reactions in Carcinogenesis: Connection Between Atherosclerosis, Ageing, Carcinogenesis and Mutagenesis

The majority of tumours have been shown to be related to environmental factors [603]; therefore, much effort has been spent on the identification of environmental carcinogens and on the clarification of their mechanism of effect. At first, it caused concern that chemical carcinogens were of such diverse structure, but later a common characteristic was found, namely the electrophilicity of their activated forms, which makes them capable of binding to DNA and other macromolecules. The majority of carcinogens have been found either to be free radicals, to be involved in free radical reactions, or to induce pathological free radical reactions (e.g. by $O_2^-\cdot$, which is an autooxidation product). The enzymes needed for activation, e.g. cytochrome P-450, may be found not only in the endoplasmic reticulum, but also in the nuclear membrane, and thus the active metabolites produced may damage the genetic material directly. There are several mechanisms which protect against the effect of chemical carcinogens: (a) prevention of the formation of active forms; (b) stimulation of the systems detoxifying carcinogens so that they cannot react with their targets; and (c) protective mechanisms acting after the interaction with the target [149, 213, 569].

Many reports have been published on the anticarcinogenic effect of antioxidants in animal experiments [25, 92, 490, 581, 582, 602]. This protective effect may be due to any of the three mechanisms. An example of the first mechanism: the formation of *N*-nitroso (-N-N-O) carcinogens can be prevented with vitamin C, vitamin E, phenol-type antioxidants (e.g. BHA) and with MTDQ-DS, a dihydro quinoline-type antioxidant (see Chap. VII Sect. 2) [47, 364, 569]. These carcinogens may reach their terminal form in the environment as well as in the organism. For this, they need an amine or amide substrate, a nitrosifying agent and the appropriate chemical milieu. Antioxidants compete with the amines and amides for the nitrosating agent (e.g. nitrite) and in this way prevent the formation of *N*-nitroso carcinogens [47, 364]. When acting by the second mechanism, antioxidants inhibit the function-related deactivation of enzyme systems mediating the metabolism of xenobiotics (e.g. the microsomal monooxidase system, epoxide hydrolase, UDP-glucuronyl transferase, gluthathione-*S*-transferase). As has been mentioned in connection with cytochrome P-450, this deactivation may be partly or completely due to LPO elicited by ROIs released during enzyme activity. The increased activity of the enzymes leads to detoxification and thus to anticarcinogenic action, but may occasionally activate the carcinogenic substance. The protective effect of vitamin A and corticosteroids is an example of the third mechanism [569]. The anticarcinogenic effect of diets containing a large proportion of vegetables and fruit, which has been proved in epidemiological studies, is due to their high vitamin A and C contents or to ingredients which can induce glutathione transferase participating in the metabolism of carcinogens [289, 349, 569].

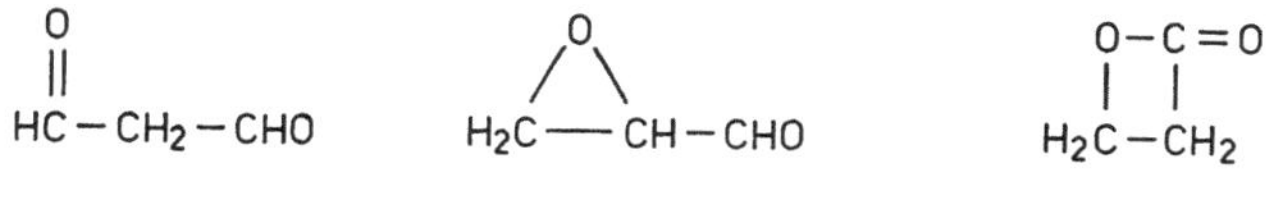

Fig. 34. Similarity between the structure of malondialdehyde and two known carcinogens. (After Shamberger [490])

Free radical mechanisms also play a role in photocarcinogenesis (see Sect. 26). The carcinogenic role of diets with high PUFA and low antioxidant content has been demonstrated [232, 233, 426]. Anthracycline antitumour antibiotics, which act by releasing free radicals, may induce the development of secondary carcinomas. Tumour frequency increases with age. An interesting correlation has been found between ageing and atherosclerosis on the one hand, and carcinogenesis and mutagenesis on the other. LPO is known to play a major role in these processes. Malondialdehyde (MDA), which has been shown to have a carcinogenic and mutagenic effect is one of the terminal products of LPO [385]. The structure of MDA resembles that of two isomers: glycide aldehyde and β-propiolactone, the carcinogenic effect of which has been proved (Fig. 34). It has been shown that MDA is ultimately responsible for their carcinogenicity. Thus during LPO occurring in the organism in connection with for example ageing and atherosclerosis, a carcinogenic and mutagenic substance is produced [490].

The theory of membrane-mediated chromosomal damage put forward by Cerutti and co-workers is based on similar considerations. They studied the mechanisms of chromosome aberrations induced with phorbol myristate acetate (PMA), a tumour-inducing croton oil derivative, and found that it acts indirectly, not by binding to DNA. It activates the respiratory burst of phagocytes and the arachidonic acid cascade, during which ROIs are produced and LPO is initiated in the course of the $PGG_2 \rightarrow PGH_2$ conversion. The metastable lipid peroxides and some stable aldehyde terminal products of LPO, like MDA and 4-hydroxy-2,3-trans-nonenal are diffusible clastogenic factors and can cause genetic damage in situ or by diffusing into other cells and tissues. In PMA-induced chromosome aberrations antioxidants and scavengers (CuZnSOD, SeGP mannitol, BHT, BHA) showed an anticlastogenic effect, with the exceptions of the singlet oxygen scavenger dimethylfuran and CAT. Inhibitors acting at different levels of arachidonic acid metabolism were also found to be anticlastogenic (the phospholipase A_2 inhibitor fluocinolone acetonide, the cyclooxygenase inhibitors indomethacin and flufenamic acid and the lipoxygenase inhibitor BN 1015 strongly; the lipoxygenase inhibitors BN 1048 and nordihydroguairetinic acid, 5,8,11,14-eicosatetraynoic acid inhibiting both cyclooxygenase and lipoxygenase; and the thromboxane synthetase inhibitor imidazole only moderately).

Other phorbol esters, mezerein, which has a similar structure, and indol alkaloid promoters (e.g. teleocidine), which are structurally different, act by the same mechanism. The protease inhibitors with strong antipromoter properties (e.g. soybean inhibitor) and also the vitamin A analogues inhibit $O_2^-\cdot$ production by phagocytes (neutrophils, alveolar macrophages) stimulated by promoters. The peroxisome proliferator compounds (clofibrate, ciprofibrate, naphenopine, di-(2-ethy-hexyl)-phthalate) cause indirect chromosomal damage similarly to membrane active agents. They are hepatocarcinogens and promoters. Their carcinogenic and mutagenic effect is due to H_2O_2 released as a result of peroxisome hyperfunction. It has been supposed that other membrane active substances, chronic inflammation and endogenous pathological free radical reactions may cause chromosome aberrations in the same way. These aberrations may play a decisive role in the initiation and promotion of tumours [70, 135, 138, 164, 306, 450, 553, 602].

The theory has been substantiated by the finding that clastogenic factor is probably composed of free arachidonic acid, aldehydes and lipid hydroperoxides [70]. In addition, lipid peroxides themselves can activate chemical carcinogens, e.g. *N*-hydroxy-*N*-acilyl-2-aminofluorene, via free radical mechanisms. Thus the stable, active substances formed on membrane damage constitute a link between atherosclerosis and ageing on the one hand, and carcinogenesis and mutagenesis, on the other [431]. Univalent leakage occurring during the activity of the mitochondrial respiratory chain may be another such link. Oberley and Buettner formulated an interesting hypothesis concerning the role of SOD in carcinogenesis. These authors and other teams invariably found the absence, or a strong decrease, of mitochondrial MnSOD in mostly animal tumour cell lines. In addition, a decreased CuZnSOD level was demonstrated in tumours, although not invariably, and the levels of GP and CAT were also low in general. The mitochondrial $O_2^-\cdot$ production was normal or increased in the tumour cells. It has been asked whether the MnSOD change was the cause or consequence of carcinogenesis. To answer this question it is necessary to look at the results to be expected from the decrease or absence of MnSOD (Fig. 35).

Increased intracellular $O_2^-\cdot$ concentration leads to the inhibition of SH enzymes, e.g. adenylate cyclase, resulting in a decrease in cAMP level. Free radicals, however, have been shown to activate guanylate cyclase, thereby increasing the cellular cGMP level [368, 407]. Cyclic nucleotides play an important part in the regulation of mitosis and thus it is understandable that a shift in their balance enhances uncontrolled proliferation characteristic of malignant cells. Since iron is a cofactor in a number of enzymatic reactions, the iron status is important from the point of view of carcinogenesis. For instance, the iron chelate picolinic acid shows a selective tumour cell toxicity.

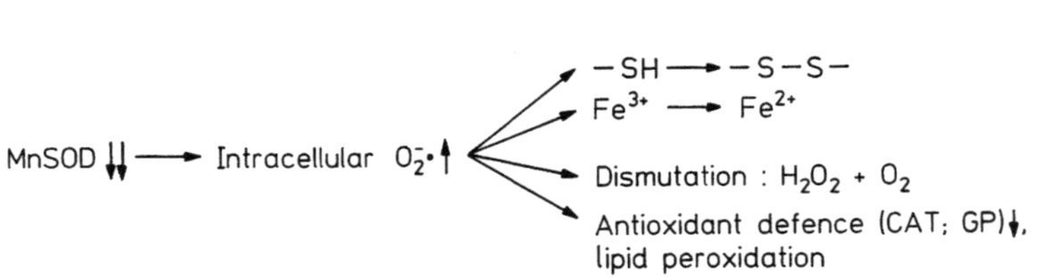

Fig. 35. The possible consequences of decrease in the MnSOD level. CAT catalase; GP glutathione peroxidase. (After Oberley and Buettner [407])

$^1O_2$ produced during dismutation; this and the $\cdot OH$ formed as a result of the interaction of H_2O_2, also formed during dismutation, with $O_2^-\cdot$ are more potent reactive intermediates, which have an increased damaging effect. LPO, which increases as a result of the decrease in the level of protective enzymes (CAT, GP) inhibits the activity of phospholipid-dependent enzymes (cytochrome oxidase, succinyl dehydrogenase). Therefore, aerobic metabolism is replaced by anaerobic processes in the tumour cells. The authors use these examples to show that the decrease in MnSOD - or its absence - is not merely a consequence of carcinogenesis, but a major change, which may be the cause of carcinogenesis, because its consequences are profound and may account for several characteristics of carcinogenesis. The following therapeutic possibilities arise from Oberley's hypothesis:

1. $O_2^-\cdot$ therapy is based on the absence or decrease of MnSOD in tumour cells, which may thus be selectively damaged. $O_2^-\cdot$ and ROI production may have a part in the antitumour effect of anthracyclin compounds (doxorubicin, daunomycin), which produce $O_2^-\cdot$ during microsomal activation, of mitomycin C and of the ROI inducer bleomycin. In the case of anthracyclin compounds this mechanism is held responsible for the side effects, rather than for the antitumour effect, which is probably due to the direct association of the anthracyclin molecule to DNA and the consequent damage of the genetic material. Photoirradiation, a new tumour therapy, which is still in an experimental stage, is also based on ROI (singlet oxygen) generation. It makes use of a synthetic derivative of haematoporphyrin, an excited form of which is produced by absorption of visible light. Its surplus energy is used for the production of singlet oxygen from molecular oxygen. Other ROIs may also be produced at the same time. The haematoporphyrin derivative selectively accumulates in tumour tissue, which is an advantage from the therapeutic point of view. This property may also be used for defining the localization of the tumour. It has been found that the inhibition of mitochondrial succinate dehydrogenase and cytochrome *c* oxidase and of the cytosol-contained pyruvate kinase (and, to a lesser extent, lactate dehydrogenase) contribute to the effect of photoirradiation. This observation supports to some extent the results of other authors who found ROIs to inhibit mitochondrial succinate dehydrogenase and cytochrome oxidase. The inhibition of cytoplasmic enzymes catalysing the terminal steps of glycolysis makes it difficult for the substrates to enter the citrate cycle and thus aerobic respiration is increasingly impeded [257, 296].

2. Diethyldithiocarbamate (DDC, a disulfiram derivative) therapy. DDC inhibits CuZnSOD, but does not inhibit MnSOD. It follows that the normal cells containing MnSOD survive the therapy, while the tumour cells containing little or no MnSOD will be selectively killed. There are experimental data supporting the therapeutic effect of DDC, but because it acts by binding copper, it may also have a toxic effect on other cuproenzymes.

Oberley's hypothesis reconciles the earlier Warburg hypothesis, now looked upon as outdated, with the modern DNA theory of carcinogenesis. According to the former, carcinogenesis is due to irreversible damage of respiration leading to the development of anaerobic metabolism, while the latter maintains that cancer is the expression of a change in DNA. This latter theory is supported by the observation that most activated carcinogens are at the same time mutagenic. It fails, however, to account for the lesion of the gene coding for, or the region regulating, MnSOD synthesis, which might be caused by exposure to background radiation or peroxidative processes (inflammation, atherosclerosis, ageing). Alternatively, MnSOD may be inactivated directly by ROIs accumulated in the mitochondria for some reason [407].

In trisomy 21 and in Dubin–Johnson syndrome characterized by increased tumour frequency, a lower MnSOD level has been found. Dimethylsulphoxide (DMSO, an effective ·OH scavenger), when added to tumour cells, caused the morphological differentiation of erythroleukaemia cells, promyelocytic leukaemia cells and neuroblastoma cells. SOD administration has been reported to prolong the survival of tumorous animals in a few cases. Naturally, antioxidants cannot be expected to reverse carcinogenesis completely; they cannot influence the irreversible changes [407]. Also, SOD cannot pass through the cell membrane, which difficulty can be eliminated by enclosing SOD in liposomes [366]. Oberley's hypothesis has not been supported by the experiments of Marklund and co-workers. They studied normal and neoplastic human cell lines and found no difference in the MnSOD level between normal and malignant cells. CuZnSOD was slightly lower in the malignant cells. No difference could be demonstrated in the levels of other enzymes (CAT, GP) either. The reason for the differences between the results of the two teams may lie in the fact that while Oberley used a smaller number of animal cell lines, Marklund investigated a high number of mostly human tumour lines. Whereas in primates MnSOD is also found in the cytosol, in other species it occurs only in the mitochondria [342].

The results of Tisdale and co-workers resemble those described in connection with Oberley's hypothesis. They investigated animal tumours and found lower SOD and CAT levels in most of them, but did not find total MnSOD and CuZnSOD deficiency [541]. Capel and Thornley found elevated SOD activity in the blood of mice with Lewis pulmonary carcinoma. In the tumour tissue SOD activity was significantly lower than in normal pulmonary tissue. An increased lipid peroxide level was found in the liver, and an even higher level in the tumour tissue [62, 63]. The experiments of Fernandez-Pol and co-workers illustrate the role of free radical reactions in carcinogenesis and support Oberley's hypothesis. They studied the effect of paraquat, a herbicide inducing intracellular $O_2^{-}\cdot$ production, on normal rat kidney cells and on virus-transformed cells. A selective cytotoxic effect on the transformed renal cell cultures was found, whereas the normal cells were resistant to the effect of the herbicide. At the same time, a normal SOD level was found in the normal cells, which increased during treatment, while in the transformed cells the SOD level was low and did not change under the effect of the treatment. Thus a good correlation was found between the response of the normal and transformed cells to the cytotoxic effect of paraquat and their ability to increase their SOD level in response to treatment. A revertant strain was found in the culture transformed with Kirsten sarcoma virus, which showed a much higher

SOD level than normal cells after paraquat treatment. This strain not only lost the characteristics of the transformed cells, but it also failed to grow on soft agar (the ability to grow on soft agar is regarded as an in vitro property correlating with in vivo tumorigenesis) and could not be distinguished morphologically from the normal cells. However, if paraquat (inducing high SOD activity) was left out of the medium, the cells regained their transformed properties [162].

Oberley and co-workers have recently expanded their theory [333, 408]. They maintain that carcinogenesis is a two-step process. In the first step, corresponding to initiation, the cells become "immortal". "Immortality" means that at the division of the initiator-damaged stem cell, two stem cells are produced, unlike normal cells which divide into a stem cell and another cell committed to differentiation.

A single application of the initiator is sufficient. The electrophilic initiator binds covalently to DNA. During the second step, called promotion, cell proliferation becomes uncontrolled. The promoter has to be applied repeatedly after initiation. The process is reversible at first, becoming irreversible only later. Promoters are, as a rule, not mutagenic and they do not bind to DNA. Many promoters are membrane active compounds (e.g. phorbol esters). Oberley and co-workers did not find MnSOD activity in the "immortal" cell line, while CuZnSOD activity was normal and could be induced with triiodothyronine (T_3). (T_3 induces SOD by increasing $O_2^{\bar{\cdot}}$ production by an unknown mechanism.) In the malignantly transformed cell line, no MnSOD activity was found and the CuZnSOD activity, too, was very low and could not be induced with T_3. The conclusion was drawn that "immortality", which, in fact, means the inhibition of differentiation, is associated with the decrease of MnSOD activity and the loss of its inducibility, while $O_2^{\bar{\cdot}}$ production is unaffected. There are data indicating that the levels of MnSOD and CAT increase with differentiation, it may be, therefore, supposed that the changes found in MnSOD activity inhibit differentiation. Retrovirus oncogenes are supposed to play a role in the transformation (see later discussion). The decrease in the activity of CuZnSOD and the loss of its inducibility with T_3 may make it suitable for the demonstration of transformation. The connection between the changes in CuZnSOD activity and transformation is unclear. Further studies are needed in support of this theory.

Although the majority of the reports describe the anticarcinogenic, mutagenic and radioprotective effects of antioxidants, lately also their tumour-promoting action has been reported. Other unfavourable effects like radiosensitization, increased mutagenic activity and the enhancement of the toxicity of some chemicals have been described [289, 333, 398, 408]. These conflicting experimental data must have some other, deeper, cause than differences in experimental conditions, i.e. in the properties of the systems studied. This is also suggested by the results of studies dealing with the role of fatty acids and prostaglandins derived from essential fatty acids in the control of mitosis. Both fatty acids and prostaglandins derived from essential fatty acids in the control of mitosis. Both fatty acids and prostaglandins have a paradoxical effect on mitosis. Lipid peroxides and 4-hydroxyalkenals, the products of LPO, have a strong mitosis-inhibiting effect, which is due to their inhibitory effect on DNA synthesis. In low concentration PUFAs may increase mitosis because the members of the n-6 or linoleic acid family are PG precursors, because in low concentration PGs enhance mitosis, and also because the amount of inhibitory LPO products formed is smaller. Higher concentrations of PUFAs and PGs inhibit mitosis because greater amounts of inhibitory lipid peroxides and other LPO products are formed.

The effect of pathological free radical reactions on the balance of cyclic nucleotides is also involved in the control of mitosis. It has been described (see Chap. III, Sect. 4.1) that different concentrations of antioxidants affect arachidonic acid metabolism differently, depending on the degree of their decreasing effect on the level of lipid hydroperoxides. High concentrations of antioxidants, although able to prevent tumour initiation by pathological free radical reactions, enhance promotion once initiation has taken place, by decreasing the level of the LPO products inhibiting mitosis. Because of the low level of lipid hydroperoxides, PG synthesis also decreases and low PG levels stimulate mitosis. Low antioxidant concentrations provide a weaker protection against initiation, but they will not decrease the level of lipid peroxides inhibiting mitosis and, as a result of increased PG synthesis, the inhibition of mitosis dominates. The paradoxical effects of antioxidants can be at least partially explained in this way [87, 381].

Various fractions of tumour tissue have been investigated by several teams and a decreased LPO has been found in tumour tissue [44, 55, 193, 436]. This observation seemingly contradicts the findings described so far: since pathological free radical reactions have a part in carcinogenesis (which is a well-founded hypothesis), one might expect to find increased LPO in the tumour tissue. It has been attempted to find the reason for the decreased LPO. Decreased PUFA content, lower cytochrome P-450 concentration and higher total antioxidant capacity have been described. Decreased PUFA content decreases LPO because of a state of substrate deficiency. Cytochrome P-450 activity may contribute to the induction of LPO (see Chap. III, Sect. 3), therefore, its falling concentration may be related to the decrease in LPO [44, 55, 193, 436].

The dynamics of tumour formation in connection with free radical reactions has been studied by ESR spectroscopy. In one study the concentration of free radicals was found to increase from initiation till the half-time of tumour growth; it then decreased in the second half and increased again during metastasis formation [132]. In another, more detailed, study, different doses of benzopyrene were given to rats intratracheally, and the concentration of the non-heme iron-sulphur-containing proteins of the mitochondrial electron transport chain, free radical concentration in the mitochondrial electron transport system, and the cytochrome P-450 content of the hepatic microsomal fraction were measured. During carcinogen administration the concentration of the iron-sulphur-containing proteins increased at first, then it dropped below the control level, while the mitochondrial free radical concentration increased. The hepatic cytochrome P-450 level decreased continuously during carcinogen administration. The carcinogen was discontinued as soon as histologically demonstrable premalignant changes were identified in the lung. At this phase the concentration of the iron-sulphur-containing proteins increased, free radical concentration was somewhat higher than in the controls, and also the cytochrome P-450 level increased. All three parameters decreased significantly in the phase of definitive malignant tumour growth [597].

The contradiction of the postulated carcinogenic effect of free radical reactions and decreased LPO found in tumours can be reconciled by supposing that during initiation the balance of the free radical reactions and the protective mechanisms is shifted in favour of the free radicals, although increased LPO induces the compensatory increase of the total antioxidant capacity. Increased LPO leads to the depletion of PUFA substrates and destruction of cytochrome P-450. Because of lack of substrates LPO decreases significantly and the decrease in the level of lipid peroxides inhibiting cell division results in cell proliferation. This is the stage of promotion when the tumour starts growing uninhibited. Further studies are needed for the verification or refutation of this hypothesis, although it has experimental foundation.

A histochemical method has been developed for the demonstration of preneoplastic cells. It is based on the demonstration of decreased LPO in tumour cell membranes, by the detection of protein-bound alkenals with Schiff reagent. The cell groups demonstrated by this method show an almost perfect match to the preneoplastic changes indicated by γ-glutamyltransferase positivity [30].

The opposite effects of PGs and LPO products on cell division may, in theory, be utilized in influencing tumour growth by dietetic measures. To prevent cell proliferation the amount of dietary n-6 fatty acids, which are the precursors of stimulating PGs ought to be decreased. Instead the amount of n-9 and n-3 fatty acids, which are not PG precursors, but give rise to inhibitory LPO products, ought to be increased [87]. Antioxidant therapy of tumour patients cannot be recommended because of the twofold action described and their possible harmful effects. The decreased LPO found in tumours and the elevated antioxidant capacity described by some authors contradict Oberley's theory and other results we have described. In the light of recent data Oberley's hypothesis might be true only for the early stage of carcinogenesis.

Another interesting result should be mentioned which might help to explain the much debated difference between male and female life expectancy. The basal metabolism of women is known to be 10% lower than that of men. Mitochondrial univalent leakage is proportional to basal metabolism. If it has a role in ageing and carcinogenesis, this might account for the higher average life span and somewhat lower tumour mortality of woman [549].

Finally the role of free radical reactions in carcinogenesis must be investigated in the light of the widely accepted cellular oncogene theory of carcinogenesis. In the normal vertebrate cell DNA sequences homologous with the oncogenes responsible for the tumourigenicity of RNA tumour viruses have been found. These cellular oncogenes are normally inactive, and their transcription is very limited. It seems that the initiation of carcinogenesis is due to the activation of cellular oncogenes. The first step is the integration of the retrovirus into the genome; the retrovirus then enhances the transcription of the oncogene either by a mutation which activates a function resembling the bacterial SOS repair (see Sect. V 11), or by a mutation affecting the cellular oncogene or a regulator region. The oncogene may also be activated by being translocated to a region characterized by active transcription.

Carcinogenesis is considered a multistep process. Oncogenic activation induces transformation only if the cell is in a state of proliferation. The agents inducing this state are called promoters. Almost all the known initiators and promoters act by free radical mechanisms. Initiation and promotion cannot be sharply delimited. For instance, breakage of the DNA chain may induce both. Chain breakage and base modification leading to mutation are typical of free radical induced damage. Free radical reactions may also enhance promotion by inducing ornithine decarboxylase. This enzyme is needed for the synthesis of polyamides participating in the regulation of DNA synthesis and cell proliferation. Induction of ornithine decarboxylase by PMA is inhibited by SOD, CAT, mannitol and SOD + CAT, either via the suppression of the inactivation of preformed ornithine decarboxylase by active oxygen (SH groups are needed for ornithine decarboxylase activity), or by inhibiting induction at genome level. Thus free radicals play an important role in both initiation and promotion [39, 180, 295, 301, 330, 461, 462, 485, 602].

Irradiation, inflammation, exposure to sunlight, some chemicals, ageing and anthracycline antitumour agents may have a role in carcinogenesis. Increased tumour frequency has been found in diseases associated with repair deficiency and in autoimmune diseases. All these states have a common factor: a postulated connection with free radical events. Pathological free radical reactions may thus participate in initiation and promotion by acting directly on DNA or indirectly by changing the delicate equilibrium existing between LPO products controlling cell division, cyclic nucleotides and arachidonic acid metabolism. Depending on other factors that may influence this balance, antioxidants may either suppress or stimulate promotion; the data available so far suggest their inhibitory role in initiation.

10.1 Summary

The majority of tumours are caused by environmental factors. Although environmental chemical carcinogens are of diverse structure, they act almost without exception via free radical mechanisms: they are either free radicals themselves, or are converted to free radicals when activated, or they induce pathological free radical reactions indirectly. Three types of anticarcinogenic protective mechanisms are known: (a) prevention of the formation of active forms; (b) stimulation of the carcinogen detoxifying systems; and (c) mechanisms acting after interaction with the target. Photocarcinogenesis, diets with high PUFA and low antioxidant contents, the carcinogenic effect of some recurrent inflammations and the effect of anthracycline antitumour antibiotics causing secondary tumours all have a free radical component in their mechanism of effect. The anticarcinogenic effect of antioxidants has been proved repeatedly. Antioxidants may act via any one of these protective mechanisms.

There is an interesting connection between ageing, atherosclerosis, carcinogenesis and mutagenesis. Endogenous carcinogens and mutagens (e.g. malondialdehyde) are produced in ageing and atherosclerosis as a result of increased LPO. According to Cerutti's theory of membrane-mediated chromosomal damage, membrane active substances activate arachidonic acid metabolism (e.g. by the free radical induced activation of phospholipase A_2), during which ROIs are released. Some stable, diffusible products of LPO can damage the genetic material, thus playing a role in the initiation and promotion of tumours. Some chemicals can also cause chromosomal damage – similiarly to membrane active substances – in an indirect way be activating the respiratory burst of phagocytes and by inducing peroxisome proliferation leading to LPO. This theory has been supported by experimental evidence, but has not been fully verified. Mitochondrial univalent leakage and the disturbed balance of the protective mechanisms acting against it may constitute another link between ageing, atherosclerosis, carcinogenesis and mutagenesis.

Oberley and co-workers proposed an attractive hypothesis, according to which the balance is upset by a lesion of the gene coding for MnSOD, or of the regulator region controlling the MnSOD level. In a more recent version of their theory they maintained that tumour initiation is basically an inhibition of differentiation caused by the decreased activity and loss of inducibility of MnSOD, which is normally highly active during differentiation. The activity of CuZnSOD is, however, unaffected. Promotion, they believe, is equal to the loss of control of cell proliferation. Decreased MnSOD activity, resistant to induction, was also found in transformed cells after promotion, but CuZnSOD activity also decreased and the enzyme could not be induced. Other authors failed to demonstrate this strong decrease or absence of MnSOD activity.

Lately antioxidants have been reported to have a tumour promoter and mutagenic effect and radiosensitizing ability. These opposing effects of antioxidants (carcinogen-anticarcinogen, mutagen-antimutagen, radioprotective-radiosensitizing) may be accounted for by the dose dependence of their effect, influenced also by other endogenous factors (on the levels of prostaglandin and lipid peroxides, which have a role in the control of cell division). In low concentrations, prostaglandins stimulate mitosis, while in higher concentrations they inhibit it owing to the action of inhibitory lipid peroxides produced during PG synthesis. PUFAs, too, stimulate mitosis in low concentrations, since they are PG precursors. In high concentrations, however, when they produce larger amounts of lipid peroxides, they have an antimitotic effect. In tumours there is decreased LPO, which may be due to intracellular PUFA deficiency, cytochrome P-450 deficiency, or perhaps to a supernormal total antioxidant protection. It may be that initiation is the result of increased LPO which is due to an early prooxidant state. The tissues respond with compensatory mechanisms acting against LPO, but eventually LPO overcomes these, causes PUFA and cytochrome P-450 deficiency, which decreases LPO in the tumour tissue owing to substrate deficiency and a decreased production of ROIs which are the source of LPO. From the findings described the conclusion may be drawn that a shift in the balance between ROIs and the protective systems acting against them must play an important role in carcinogenesis. Pathological free radical reactions are involved in both the initiation and promotion of carcinogenesis.

11 Free Radical Reactions and Mutagenesis

Free radical reactions can induce mutation of DNA either directly or indirectly by a mechanism resembling the SOS repair system of bacteria. The activation of this system is induced by DNA lesions (direct mutations), but this is a repair system "liable to error", producing secondary mutations [270].

Free radicals may have a decisive role in the random mutations caused by background radiation. Mutagenesis, however, is not always harmful, it may also serve the purposes of evolution. As regards the connection between carcinogenesis and mutagenesis, Totter found that tumour is the price paid by the individual for the survival of the species as a result of useful mutations [549].

We have already mentioned the syndromes characterized by repair deficiency and that free radical reactions may lie behind these. The repair deficiency may be secondary to DNA damage caused by ROIs. The mutagenic effect of ROIs on bacteria has been proved experimentally, and a similar effect of activated phagocytes has been found in cell cultures [377, 385, 579].

The antimutagenic effect of antioxidants has been repeatedly described. However, the results of the highly sensitive cytogenic tests have not been without contradictions. For instance, ethoxyquin did not inhibit sister chromatid exchange induced by cyclophosphamide, whereas it was effective in the micronucleus test (micronuclei correspond to chromosome fragments) and chromosome aberration test. Three types of antimutagens are distinguished depending on their mechanism of effect:

1. Desmutagens, which inactivate or kill mutagens
2. Inhibitors of the metabolism of substances needing metabolic activation for their mutagenic action
3. Antimutagens proper, which inhibit the mutagenic effect at the cellular level

Ethoxyquin belongs to the third type, because if it were of type 1 or 2, it would inhibit sister chromatid exchange, since cyclophosphamide is a mutagen undergoing metabolic activation. Other antioxidants may also belong to the third type of antimutagens [101, 274, 454, 510].

The mutagenic effect of antioxidants has also been described [289, 378]. One of the possible explanations of these contradictory data was given in Sect. 10.

11.1 Summary

The mutagenic effect of ROIs has been proved. They may act either directly or by activating a mechanism resembling the SOS repair system of bacteria, which is "liable to error".

12 Free Radical Mechanisms in the Effect of Drugs and Chemicals

Drugs may be converted to reactive intermediates through oxidation or interaction with the body's enzyme systems. It seems probable that several of these chemically reactive intermediates are present as free radicals. For example dialuric acid, 6-OHDA and 7,8-dihydroxychlorpromazine easily undergo autooxidation to radical intermediates, probably to semiquinone radicals. ROI production is part of this autooxidation process [57, 247, 554].

UV light can photoactivate sulphanyl amide and 4-aminobenzoic acid, which is one of the components of photoprotective products. As a result, free radical intermediates are formed, which are probably responsible for the phototoxic effect of these agents. Long-wavelength UV light activates psoralens to radical intermediates, which form photoadducts with the pyrimidine bases in psoriatic epidermal cells and have a beneficial effect in psoriasis. However, the reaction may predispose to melanoma and squamous cell carcinoma [345, 554].

These mechanisms are nonenzymatic, but the activation of most chemical agents is enzyme mediated. The best known example is that of carbon tetrachloride CCl_4, the activation of which in the endoplasmic reticulum is mediated by mixed-function oxidase enzymes (see Sect. 15). These enzymes have been shown to occur also in the nuclear membrane, in addition to the endoplasmic reticulum. The reactive intermediates they produce are thus a direct hazard to DNA and probably play a role in the activation of anthracyclin antitumour antibiotics [176].

In addition to the enzymes of the endoplasmic reticulum, mitochondria and the nuclear membrane, cytoplasmic enzymes may also activate xenobiotics to radical intermediates. For example xanthine oxidase is capable of the one-electron reduction of doxorubicin, nitrofurantoin, misonidazole and paraquat. $O_2^{-}\cdot$ generated during xanthine oxidase activity converts 2-hydroxyoestrogens and α-methyldopa to reactive intermediates - probably to quinone and/or semiquinone, which can then bind covalently to proteins. α-Methyldopa may also be activated in the cytochrome P-450 system and can induce ROI production. Free radical events may have a part in the hepatotoxic and Coombs-positive haemolytic anaemia-inducing side effects of α-methyldopa. Cytoplasmic diaphorases facilitate the redox conversion of paraquat and of the carcinogenic benzpyrene quinones, as a result of which ROIs are generated [57, 543]. Also the PG synthetase system may participate in the activation of benzpyrenes and other xenobiotics, e.g. diethylstilboestrol and nitrofuran. Cooxidation of these substrates takes place in association with the oxidation of arachidonic acid to PG by PG synthetase [57, 446].

Activation of chemicals may be due to interaction with another chemical, reactive oxygen produced during the redox conversion of which may induce the conversion of the agent to a reactive form. Such interaction has been described between paraquat and dopa, or between mitomycin C and bleomycin leading to the potentiation of the DNA-damaging and cytotoxic effect of mitomycin C [554]. Bleomycin may act as a bound oxidant, which means that an activated molecule or complex is formed, which does not leave the site of its generation [554, 555, 556].

Paraquat (Gramoxone) is a herbicide which may cause severe, frequently lethal, poisoning. It attacks the lung in the first place, causing cyanosis and pulmonary oedema. It is converted to a reactive radical (bipyridinium cation) by the mediation of NADPH cytochrome P-450 reductase. The radical reacts with molecular oxygen producing $O_2^-\cdot$. SOD treatment has been found beneficial in paraquat poisoning [5, 57, 241, 242, 554, 587].

The bactericidal, mutagenic, carcinogenic and pulmonary-damaging effect of several nitro compounds (e.g. nitrofurantoin, nifurtimox, metronidazole chloramphenicol) is due to the nitro radical anion $RNO_2^-\cdot$ produced during activation mediated by NADPH cytochrome P-450 reductase. The process of free radical formation is shown in Fig. 36. Three reactions of $RNO_2^-\cdot$ are possible: (a) $RNO_2^-\cdot$ may react with itself, when no radical is produced (detoxification); (b) if O_2 is present, the reaction leading to $O_2^-\cdot$ formation dominates; (c) $RNO_2^-\cdot$ can also bind covalently to macromolecules. The second and third of these reactions may be responsible for such phenomena as the bactericidal effect of nitrofurantoin and other nitro compounds, or for nitrofurantoin-induced hypersensitivity pneumonitis and pulmonary fibrosis. The morphological picture of this pulmonary fibrosis closely resembles that seen in paraquat poisoning. The second and third of these reactions may also be involved in the radiosensitizing effect of metronidazole [57, 348, 554, 367].

Quinone-type antitumour agents (doxorubicin, daunomycin, rubidazone, mitomycin C and streptonigrin) are converted to semiquinone radicals by electron transfer from the microsomal and nuclear membrane electron chain. The conversion is mediated by NADPH cytochrome P-450 reductase. The semiquinone radicals, owing to their strong affinity and selective binding to nucleic acids, may act as "site-specific free radicals". However, the role in antitumour activity of this microsomal activation to semiquinone radicals has not been proved, it may also be that the antitumour agents mentioned become intercalated in the duplex structure of the DNA molecule and act directly without any free radical event. Semiquinone radicals formed during microsomal activation react with oxygen to form $O_2^-\cdot$, and the ROIs produced and the LPO these elicit are held responsible for the side effects. Both mechanisms of the antitumour effect of these quinoline-type agents may play a role in their secondary carcinogenic effect.

Cardiotoxicity is the most important side effect of doxorubicin and daunomycin, although their hepatotoxic effect may also be considerable. The relatively weak antioxidant defence of the heart is responsible for the cardiotoxic

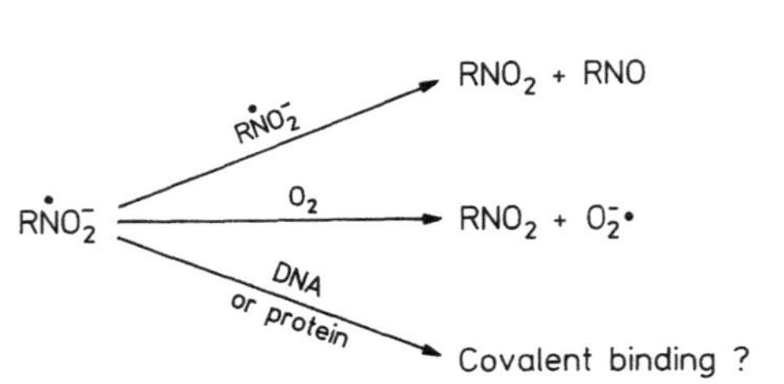

Fig. 36. Proposed chemical reactions of nitroaromatic radical anions in vitro. (After Mason and Josephy [348])

effect. The cytosol SOD level in the heart is only 80% of that in the liver. Practically no CAT is found in the myocardium, whereas it is abundant in the liver. Glutathione peroxidase contained by the mitochondrial matrix is absent from the heart and is abundant in the liver. The defence of the heart against ROIs relies on GP activity in the cytosol - this is about the same in the liver. This accounts not only for the primarily cardiotoxic side effects of doxorubicin, but also for the cardiac damaging effect of selenium deficiency (selenium is needed for the action of SeGP) [17, 127, 320, 335, 363, 437, 438, 526, 595].

It has been attempted to mitigate these side effects of anthracycline which are brought about by the indirect mechanism, while preserving the antitumour effect. For this the following possibilities are available:

1. Administration of antioxidants preventing LPO induced by ROIs, which are produced from the semiquinone radicals during microsomal activation. Antioxidants do not interfere with the antitumour effect, irrespective of whether it is due to intercalation or to the direct effect of semiquinone radicals, since they probably have no effect on the intercalation and are not likely to neutralize the semiquinone radical produced in the nuclear membrane. It is believed that they cannot neutralize the semiquinone radical because the dose is lower than that required for the inhibition of secondary LPO, and also because their physicochemical properties are unlikely to enable them to reach, in sufficient concentrations, the site of production in the nuclear membrane of the semiquinone radical, which may damage DNA directly [320, 335]. Our own results also support the LPO mechanism of the side effects of anthracycline, and the usefulness of antioxidant treatment. Increased LPO was found in association with the hepatoxicity of doxorubicin. It could be suppressed by the administration of MTDQ-DS, a dihydroquinoline-type antioxidant. At the same time, the hepatic function and the morphological picture improved [86, 563]. The effect of vitamin E on cardiotoxicity was also tested. The antioxidant did not decrease the antitumour effect, but the results in connection with its effect on cardiotoxicity were controversial. Favourable effects were registered in acute doxorubicin cardiotoxicity, while in chronic cases vitamin E was ineffective. This was probably due to the fact that vitamin E is part of a complex antioxidant system, and in chronic experiments its effect was prevented by the depletion of the GP activity [316, 390]. The antitumour action was unaffected by antioxidant treatment.

2. Inhibition of microsomal activation. The microsomal activation of doxorubicin could be competitively inhibited with methylene blue, which is also reduced by NADPH cytochrome P-450 reductase. Since methylene blue is nontoxic and did not interfere with the antitumour action of doxorubicin in the experiments, its clinical testing has been suggested [266].

3. Administration of anthracycline derivatives with modified structure, which are not, or are only slightly, activated in the microsomes, e.g. 4-deoxy-doxorubicin, 5-iminodaunomycin and (*cis*-7,9-dihydroxy)-9-acetyl-6,7,9,11-tetrahydroxyxantho(2,3)-tetraline. These drugs are not cardiotoxic, or at least their cardiotoxic effect is much weaker than that of doxorubicin, while they are potent antitumour agents [118, 334, 437, 438]. The antitumour and pulmonary fibrosis-inducing effects of bleomycin are due to free radical

events. Bleomycin is a special molecule, one part of which recognizes the special regions of DNA to which it is bound, while the other part binds iron. Its DNA-damaging effect is probably due to a bleomycin-Fe^{3+}-•OH complex [534, 554, 555, 556]. Procarbazine, another cytostatic, is also converted to a free radical via either microsomal activation, or cooxygenation mediated by prostaglandin synthetase. This may play a role in both the antineoplastic effect and the side effects of the drug [501].

The myelotoxic effect of benzene has long been known. This may lead to pancytopenia, aplastic anaemia, leukaemia and chromosome aberrations. Its effect is probably due to LPO, which it may induce either by generating $O_2^-\cdot$ via autooxidation, or by inhibiting heme synthesis and thus the incorporation of iron into the erythrocytes. As a result, iron accumulates in the bone marrow, which enhances LPO [297].

Free radical reactions probably have a part in the toxic effects of mercury (e.g. nephrotoxicity, central nervous symptoms). Hg^{2+} stimulated the LPO of hepatic mitochondria in rabbits and of hepatocytes in rats. This effect could be suppressed with antioxidants (N,N′-diphenyl-*p*-phenylenediamine, vitamin E, sodium, selenite, (+)-cyanidanol-3) [182, 519, 542]. The effect of $HgCl_2$ was studied in erythrocytes. It was found to cause haemolysis, which was preceded by peroxidation of erythrocyte membrane lipids. The mechanism of haemolysis was explained as follows: Hg^{2+} enters the erythrocyte cytosol and forms an Hg^{2+}-oxyhaemoglobin complex, which binds to the phospholipids of the erythrocyte membrane. The complex easily undergoes autooxidation to Hg^{2+}-methaemoglobin, during which $O_2^-\cdot$ is released. The half-life of $O_2^-\cdot$ is longer in the nonpolar environment of the membrane than in the aqueous medium, and also the catalysing effect of the membrane-bound Hg^{2+}-methaemoglobin complex increases LPO. The SOD, CAT and GP contained by erythrocytes cannot prevent LPO, because these enzymes neutralize ROIs only in the aqueous phase [456]. The toxic effects of acetaminophen (interstitial nephritis, renal failure, hepatotoxicity and because of which this drug is now rarely used) are also due to LPO. Although experimental data are available as regards its LPO-inducing effect, the mechanism of this effect is not clear [4, 165, 539, 580].

12.1 Summary

Drugs or chemicals may undergo autooxidation in the organism, or may interact with the body's enzyme systems to form reactive intermediates. Most of these intermediates are free radicals. Some important examples are:

1. Hepatotoxic carbon tetrachloride is activated to a reactive free radical by the mixed-function oxidase enzyme found in the endoplasmic reticulum and nuclear membrane.

2. α-Methyldopa is activated to reactive free radical by xanthine oxidase or cytochrome P-450. This is probably responsible for its hepatotoxicity and for its Coombs-positive haemolytic anaemia-inducing effect.

3. The herbicide paraquat is activated to free radical by NADPH cytochrome P-450 reductase and damages the lung.

4. Nitro compounds (e.g. nitrofurantoin, nifurtimox, metronidazole, chloramphenicol) are activated by cytochrome P-450 reductase to nitro radical anion $RNO_2^-\cdot$*, which forms* $O_2^-\cdot$ *in the presence of oxygen. The LPO-inducing effect of ROIs thus produced is involved in the bactericidal, mutagenic, carcinogenic and pulmonary-damaging effect of these compounds. This mechanism is responsible for nitrofurantoin-induced hypersensitivity pneumonitis and metronidazole-induced radiosensitivity.*

5. Quinone-type antineoplastic agents (doxorubicin, daunomycin, rubidazone, mitomycin C and streptonigrin) are activated to semiquinone radicals by NADPH cytochrome P-450 reductase. The direct binding of the semiquinone radical to DNA may contribute to the antitumour effect of these drugs. This has, however, not been proved, moreover it has been suggested that microsomal activation is not needed for the antitumour action, which is due rather to the intercalation of these agents in the double DNA structure. Nevertheless, the role of $O_2^-\cdot$ *produced by the semiquinone radical in the presence of* O_2 *has been shown to be responsible for the side effects of the drugs. The side effects (e.g. cardiotoxicity) are caused indirectly via ROI production and consequential LPO.*

6. Free radical events lie behind both the antineoplastic and pulmonary fibrosis-inducing effects of bleomycin.

7. Procarbazine probably undergoes microsomal activation, or prostaglandin synthetase-mediated cooxygenation to free radical and this is responsible for both its antineoplastic action and its side effects.

8. The myelotoxic effect of benzene may be due to LPO induced by ROIs which are produced during its autooxidation.

9. The LPO-inducing effect of mercury is involved in its toxicity.

10. Acetaminophen damages the kidney and liver via LPO induction.

13 Free Radical Reactions in Pulmonary Diseases

13.1 Emphysema

Air pollution and smoking play a major role in the pathogenesis of centriacinar emphysema. These factors probably give rise to free radical events. Ozone and nitrogen oxide, the main components of photochemical smog enter through the airways and cause LPO [229, 304, 351]. High concentrations of free radicals have been demonstrated in soot found in the air of cities, in the exhaust gases of cars, in burning and nonburning tobacco, and in atmospheric dust. In cigarette smoke nitrogen dioxide is formed via slow oxidation from the relatively unreactive nitric oxide present in high concentration. Nitrogen dioxide generates peroxy and alkoxy radicals from the hydrocarbons and dienes in the smoke

$$NO \xrightarrow{\text{slow initiation}} NO_2 \xrightarrow{\text{hydrocarbons, dienes}} R\cdot \xrightarrow{O_2} ROO\cdot \xrightarrow{NO} RO\cdot$$

↘ ↓ ↙

Radical-terminating reactions

It might be expected that these radicals are short-lived; however, Pryor has found that their half-life is approximately 5 min in the gas phase and they can travel a relatively long distance in cigarette smoke without a considerable decrease in their concentration. This is probably due to the development of a state of balance between the incessant formation of free radicals and the radical-terminating reactions. Thus, LPO is partly responsible for the harmful effects of smoking. Free radicals probably also play a part in the carcinogenicity of cigarette smoke. Polycyclic aromatic hydrocarbons (PAH), which constitute the main part of the tar fraction, are the major carcinogens of cigarette smoke. Their concentration is, however, not sufficiently high to account for the carcinogenic effect. $O_2^-\cdot$ may be produced during the enzymatic or spontaneous oxidation of the hydroxyquinoline rings of polyphenol which is also found in the tar fraction. This $O_2^-\cdot$ may participate in the activation of PAHs to carcinogens, while semiquinone is formed from the hydroquinone. Semiquinone may bind covalently to DNA bases, primarily to guanine, which increases the carcinogenic effect [80, 445, 446].

Disturbance of the protease-antiprotease balance is also involved in the emphysema-inducing effect of air pollutants (ozone, nitrogen dioxide) and smoking. Emphysema may occur at an early stage in hereditary α_1-antitrypsin deficiency, which is, however, very rare. Ozone, NO_2 and cigarette smoke attract alveolar macrophages and neutrophils to the site of exposure and activate them. Injury and stimulation of alveolar macrophages may be the major early effects which lead to the release of the inflammation-promoting factors of phagocytes. Macrophages have been shown to produce neutrophil chemotactic factors which attract neutrophils to the lung (normally, pulmonary tissue contains very few neutrophils). In addition, factors enhancing neutrophil adhesion and $O_2^-\cdot$ release are probably produced by the alveolar macrophages under the effect of hyperoxia. Alveolar macrophages may also be stimulated by some components of cigarette smoke, by the bacteria present and by the lymphokines of the lymphocytes attracted to the site of injury [192, 279, 355, 362, 476, 537]. It has been shown that the alveolar macrophages of smokers produce significantly more H_2O_2 compared with nonsmokers. The NO and NO_2 in gas phase smoke may interact with H_2O_2 produced by macrophages while potent $\cdot OH$ is produced.

$$NO + H_2O_2 \rightarrow HO\cdot + HONO$$
$$NO_2 + H_2O_2 \rightarrow HO\cdot + HONO_2$$

Elastase released by the phagocytes may damage the elastin of alveoli, because the ROIs released inactivate α_1-antitrypsin. Owing to its oxidant free radical content, cigarette smoke may directly inactivate α_1-antitrypsin, thus producing acquired α_1-antitrypsin deficiency [148, 185, 192, 212, 278, 279, 446, 537]. An increased coeruloplasmin level was found in the bronchoalveolar lavage fluid of smokers. This was associated with an increased serum coeruloplasmin level, nevertheless, the total antioxidant capacity of serum was lower than in the controls. This shows that coeruloplasmin is not capable of providing full protection against oxidant stress. In this localization one of the functions of coeruloplasmin may be the protection of α_1-antitrypsin against oxidants [192].

SO_2 is also a major air pollutant. The sulphite anion SO_3^{2-} and bisulphite anion HSO_3^- (its ionized forms) occur in food preservatives. SO_2 is hydrated to HSO_3^- in the lung. PG hydroperoxidase, and probably also other peroxidases, can oxidize the bisulphite anion to sulphur trioxide anion free radical $SO_3^{-}\cdot$, which can induce pathological free radical reactions [347].

Air pollutant oxidant gases may cause proneness to infections of the respiratory tract by damaging the mucociliary transport system, which is the primary defence mechanism of the lung, protecting the larger bronchi. These gases also damage the secondary protection of the areas of gas exchange provided by the alveolar macrophages, which they may also stimulate [190].

13.2 Bronchial Asthma

The role of free radical reactions has been suggested, in addition to that of allergy, in bronchial asthma and immediate hypersensitivity reactions. This has been based on the free radical activation of PG synthesis and on the fact that SRS-A is a mixture of several types of leukotriene (see Fig. 14). LTD_4 for example is 140–150 times more potent a bronchoconstrictor than histamine. The figure also provides an interesting explanation of the pathogenesis of aspirin-induced asthma: aspirin and nonsteroid anti-inflammatory agents inhibit the cyclooxygenase reaction pathway and thereby shift the balance in the direction of the lipoxygenase pathway. SRS-A, an asthma-inducing agent, is produced in this pathway. This, naturally, does not explain why aspirin does not cause asthma in healthy individuals, in whom the same shift occurs. This contradiction may be accounted for on the basis of recent experiments which have shown that hyperreactivity of the airways - a characteristic change in asthma - may also be found with products of arachidonic acid metabolism. PGD_2, $PGF_{2\alpha}$ and LTD_4 have been shown to induce stronger and more lasting airway obstruction in asthma patients than in controls.

Oxidant gases induce a hyperreactive state of the airways which is analogous to asthma [38, 203]. Aspirin-induced asthma and Quincke's oedema could be suppressed with vitamins E and C, which points to the role of free radical reactions [375]. Increased levels of hydroperoxides, MDA, conjugated dienes and Schiff bases, and increased activity of endogenous phopholipids were found in the bronchial secretion of 5- to 14-year-old bronchial asthma patients in the acute phase of the disease. The CAT and β-carotene levels decreased significantly, while peroxidase activity showed a compensatory increase. In periods of remission the peroxide level and phospholipase activity decreased [307].

In addition to the ROI effect on arachidonic acid metabolism described in Chap. III Sect. 4.1, resulting in a shift in the PGI_2/TXA_2 balance, oxidant gases and smoking may also have systemic effects which are due to their interference with the pulmonary inactivation of PGs. The lung inactivates PGs, with the exception of PGI_2, which does not bind to the carrier protein transporting PGs into the cells. Inactivation is mediated by two enzymes: 15-hydroxy-PG-dehydrogenase (PGDH) and 13,4-reductase. Oxidant gases inhibit PGDH activity and thereby upset the PGI_2/TXA_2 balance, producing TXA_2 excess [133]. Respiratory tract infections may activate phagocytes, and the ROIs released trigger the pathway mentioned and produce the spastic component causing chronic spastic bronchitis. The source of leukotrienes is not known in asthma, they may be released by mast cells or phagocytes [103, 203, 204, 214, 307, 375, 479, 576].

13.3 Acute Respiratory Distress Syndrome in Adults

In acute respiratory distress syndrome (ARDS) occurring in adults, the injury of the endothelium of pulmonary vessels is the primary lesion. As a result, vascular permeability increases and permeability-type pulmonary oedema develops. Pulmonary neutrophil sequestration is another characteristic feature of the syndrome. Pulmonary vascular resistance increases. ARDS may be associated with shock of diverse aetiology, sepsis, microembolization, hyperoxia, pancreatitis and aspiration. Activated neutrophils circulate in the blood, which, depending on the primary condition, may be due either to complement activation via an alternative pathway (see Craddock's mechanism in Sect. 3), or to activating agents released by macrophages (e.g. ARDS elicited by oxidant gases). Complement activation and the activation of alveolar macrophages result, in the production not only of neutrophil activating substances, but also of neutrophil chemotactic products. Thereby they attract neutrophils to the site of the reaction. ROIs can also produce neutrophil chemotactic factor during their interaction with a plasma precursor. It has been proved that activated neutrophils damage the endothelium via the release of proteases, ROIs and arachidonic acid derivatives. This effect is responsible for the early changes in ARDS. Neutrophil depletion and the protective effect of free radical scavengers have been proved in animal models of ARDS induced with endotoxin, microembolization, zymosan-activated plasma and oxygen radical generating systems. While there are several reports available describing the results of relevant animal experiments, the clinical data are restricted to a few cases of ARDS in which the protective effect of glucocorticoids is probably also based, at least partially, on their phospholipase A_2 inhibitory and antioxidant actions [65, 224, 275, 343, 537, 588].

13.4 Silicosis and Asbestosis

Free radicals participate in the pathogenesis of silicosis, an industrial disease causing much concern. Hungarian authors have demonstrated the haemolytic action of quartz, which was associated with increased LPO. Haemolysis could be blocked by vitamin E administration. Quartz damages type I pneumonocytes responsible for gas exchange. This, and probably also a defect of surfactant synthesis, result in hyperplasia of the surfactant-secreting type II pneumonocytes. As a result, increased amounts of surfactant are released from the multilamellar bodies of these cells. The lipids released and quartz are taken up by the macrophages, and this leads to the activation of their respiratory burst and the release of ROIs. This hypothesis was supported by the significant increase found in the levels of MDA, SOD and GSH; the latter two may represent a protective mechanism [376, 540, 604].

Asbestos is composed of hydrated silicates of fibrous structure. Diffuse pulmonary fibrosis may develop as a result of asbestos inhalation. The incidence of bronchogenic cancer, pleural and peritoneal mesothelioma is also higher in workers exposed to asbestos. Prolonged exposure probably results in the chronic accumulation of macrophages and neutrophils in the alveoli since phagocytes cannot eliminate the asbestos fibres. The cytotoxic products released by these cells (e.g. elastase, ROIs) cause chronic damage of the lung [382, 605].

13.5 Summary

Air pollutants (ozone, nitrogen dioxide, sulphur dioxide) and smoking damage the lung via free radical mechanisms. A pathogenic role has been attributed to them in centriacinar emphysema. The carcinogenic effect of smoking has been attributed in part to ROIs found in the cigarette smoke and in part to the carcinogenic substances occurring in the tar fraction. These, too, are activated by free radical reactions. The participation of free radical reactions in the pathogenesis of bronchial asthma and chronic spastic bronchitis has been suggested. Owing to the connection of free radicals with arachidonic acid metabolism, they can contribute to the production of SRS-A, a product of the lipoxygenase pathway, which can induce asthma. This, however, needs further verification. In the acute respiratory distress syndrome of adults, the cytotoxic products (e.g. ROIs, elastase) released by the activated, sequestered neutrophils play an important part. This mechanism has been proved in animal models of ARDS, while in humans verification is still lacking, only the beneficial effect of antioxidant therapy has been described. Quartz causes silicosis by inducing LPO. Inhalation of asbestos fibres causes diffuse pulmonary fibrosis and increased incidence of bronchogenic cancer and mesothelioma. This is due to the chronic accumulation of macrophages and neutrophils (since phagocytes cannot eliminate the asbestos fibres). Cytotoxic products released by alveolar macrophages and neutrophils may be responsible for the lesions.

14 Free Radical Reactions in Hepatic Diseases

The liver is one of the organs best supplied with antioxidants. This is most probably due to its detoxifying function in the course of which ROIs are released. Peroxidized lipids can be demonstrated in the bile even under normal conditions [126, 263]. Large acute doses of toxic agents or chronic exposure to such substances overpower the antioxidant defence system and cause hepatic damage.

14.1 Alcoholic Liver Damage

Three forms of alcoholic liver disease may be distinguished. These are, in order of increasing severity: steatosis, alcoholic hepatitis and alcoholic cirrhosis.

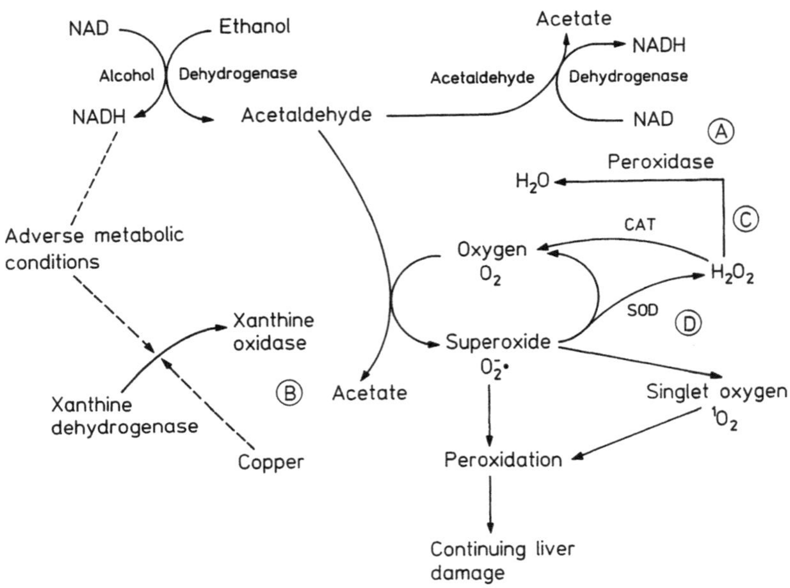

Fig. 37. Hypothetical free radical pathogenesis of alcoholic liver damage. (After Lewis [325])

LPO probably plays a causative role in each of these, although experimental evidence has been obtained only in steatosis. Decreased hepatic GSH and increased conjugated diene levels were found in biopsy specimens of the liver of patients with alcoholic liver disease [474, 493]. In rats exposed to ethanol inhalation, decreased hepatic CAT and CuZnSOD activities were found, while the MnSOD activity was normal. The changes were attributed to ROIs released during the microsomal metabolism of ethanol [457]. Both acute and chronic alcohol administration to rats and baboons produced an elevation of the level of conjugated dienes and a significant depression of GSH in the liver [492]. The mechanism of alcohol-induced LPO is not fully known. Lewis and Paton [325] suggested the role of cytoplasmic xanthine oxidase (Fig. 37).

Acetaldehyde, the first metabolite of alcohol, is metabolized not only by acetaldehyde dehydrogenase, which is its chief metabolizing enzyme, but also by xanthine oxidase. This enzyme has been shown to function in normal cells not as an oxidase, but as an NAD^+-reducing dehydrogenase (see Sect. 8). However, under unfavourable metabolic conditions such as proteolysis or ischaemia, it functions as an oxidase enzyme. Instead of hypoxanthine, acetaldehyde may be the substrate of this enzyme; it is oxidized to acetate, generating meanwhile $O_2^{-}\cdot$ from molecular oxygen.

The presence of large amounts of acetaldehyde in alcoholics (as opposed to a limited supply of purine metabolites) leads to excessive $O_2^{-}\cdot$ production [325]. The hypothesis of Lewis and Paton has been supported by experimental evidence: in rats administration of tolbutamide, which inhibits aldehyde dehydrogenase, increased the enthanol-induced depression of GSH by increasing the circulating acetaldehyde level, whereas allopurinol had an opposite effect. Acetaldehyde incubated in vitro with bovine xanthine dehydrogenase/oxidase selectively inactivated xanthine dehydrogenase, which result differed somewhat from what had been expected based on the Lewis-Paton theory. At the same time, acetaldehyde did not affect the oxidase activity of the enzyme. Accordingly, the ROIs produced by acetaldehyde are not the result of its metabolization by xanthine oxidase, but of a $D \rightarrow O$ conversion [410].

Another possibility would be that alcohol increases the activity of hepatic microsomal enzymes, including the microsomal ethanol oxidizing system. ROIs released during the activity of microsomal enzymes may play a role in the hepatic LPO-inducing effect of alcohol [418, 457, 492]. In addition, acetaldehyde can generate ROIs not only via its metabolism catalysed by xanthine oxidase as supposed by Lewis and Paton, but also during its aldehyde oxidase-mediated metabolization [457, 522]. The binding of acetaldehyde to the SH-containing GSH precursors (see Sect. 9), whereby GSH concentration decreases, is a third alternative mechanism [189].

Increased CuZnSOD activity was found in the erythrocytes of chronic alcoholics with hepatic lesions. It has been suggested that this finding may be used as a marker of ethanol consumption in addition to other tests [137].

Free radicals may have a part in the pathogenesis of haematological changes associated with alcoholic liver disease (macrocytosis, haemolytic anaemia, target cell and spur cell production). Decreased red cell PUFA, GSH and vitamin E levels and decreased pyruvate kinase activity, and a lower serum vitamin E level were found in patients with Zieve syndrome (alcoholic liver disease associated with hyperlipidaemia and haemolytic anaemia). The erythrocytes of the patients showed increased sensitivity to H_2O_2 stress in vitro. Thus, in addition to a number of factors (folic acid deficiency, decreased prothrombin level), the disturbance of the oxidant-antioxidant balance may play an important role in the development of the haematological complications of alcoholic liver disease [201].

We investigated the role of free radical reactions in the pathogenesis of chronic liver diseases. The lysosomal enzyme activity was assayed in the serum and granulocytes of such patients. LPO is known to be able to damage lysosomal membranes, causing lysosomal enzyme release. Significantly increased serum β-glucuronidase activity was found in steatosis hepatitis, chronic active hepatitis (CAH) and in cirrhotic hepatitis (CH) alike. Serum acid phosphatase activity, too, increased in each of the three conditions, but in CH the increase was not significant. A nonsignificant increase of serum cathepsin D activity was found in all three groups. The β-glucuronidase activity of granulocytes was lower in hepatopathic patients than in controls. The rate of β-glucuronidase release from granulocytes increased in patients with chronic liver disease. Thus the higher serum β-glucuronidase level originated in part from the granulocytes, although this does not account for all of the increased enzyme activity in serum.

As a result of the decreased stability of lysosomal membranes, there is an enzyme outflow from other tissues, and thus also from hepatic tissue. In CH, lysosomal membrane injury was demonstrated electron microscopically. The

β-glucuronidase level was highest in CH, and a less intensive increase was found in CAH and steatosis hepatitis. The decrease of granulocytic β-glucuronidase activity was, accordingly, strongest in CH. Thus the testing of serum and granulocytic lysosomal enzymes may be used in the diagnosis of liver diseases, similarly to the hepatic mitochondrial and microsomal enzyme assay. Moreover, these enzyme levels may furnish useful differential diagnostic information in chronic liver diseases [154, 157, 160, 543, 544]. A course of 3 months administration of (+)-cyanidanol-3(Catergen), a drug with antioxidant action, normalized the increased acid phosphatase level, significantly decreased the elevated β-glucuronidase activity and normalized the release of the enzyme from granulocytes in chronic liver disease [158]. Of the antioxidant, membrane stabilizing hepatoprotective agents, 4-amino-5-imidazolcarboxamide (AICA), Catergen, MTDQ and MTDQ-DS were found to have immunomodulator action. This, however, varied: the drugs were immune suppressive in some tests, while in others an immune stimulator action was observed [151]. Legalon, containing the membrane stabilizing and antioxidant agent silymarin, was effective in various forms of drug-induced hepatotoxicity (e.g. CCl_4, galactosamine, ethanol, thioacetamide, *Amanita phalloides* poisoning), in chronic liver diseases (chronic persisting hepatitis, chronic active hepatitis, cirrhosis), and in acute hepatitis [29, 37, 68, 98, 451].

14.2 Hepatic Damage Caused by Drugs and Chemical Agents

Carbon tetrachloride and galactosamine are well-known hepatotoxic agents. They only have a damaging effect if there is an active hepatic metabolism. CCl_4 is reduced in the hepatic endoplasmic reticulum by cytochrome P-450, the terminal oxidase of the mixed-function oxidase system. $CCl_3\cdot$ is formed by the cleavage of the CL_3C–Cl bond. This is rapidly converted to $CCl_3O_2\cdot$, a more electrophilic and more reactive species. This radical induces LPO in the endoplasmic reticulum. Thus in CCl_4-induced hepatic damage LPO occurs mainly in the microsomes. CCl_4-induced hepatic damage is regarded as the model of hepatic injury caused by xenobiotics [356, 455, 505]. As opposed to CCl_4, galactosamine is water-soluble and the radical reactions it elicits take place in the aqueous phase. In this case, too, LPO is involved in the damage, but is not its primary cause. Galactosamine-induced hepatic damage is the model of inflammatory hepatopathies (virus hepatitis) [593]. In experimental animals chronic administration of CCl_4 caused hepatoma, furnishing thereby additional proof for the role of free radical reactions in carcinogenesis [356]. Antioxidant therapy proved to be useful in the clinical treatment of acute CCl_4 poisoning: intravenous acetylcysteine minimized the subsequent hepatorenal damage [473].

Following administration of CCl_4 and galactosamine to experimental animals we found an increased MDA level and increased β-glucuronidase release (indicators of LPO-induced damage) in the liver. MTDQ decreased the hepatotoxic effect of CCl_4 significantly, as indicated by laboratory data and the morphological picture. Water-soluble MTDQ-DS diminished the hepatotoxic effect of galactosamine significantly (see Sect. 2) [150, 153, 160, 517, 545,

548]. The protective effect of Catergen was also demonstrated in vitro: it prevented the increased release of lysosomal enzymes, which is a characteristic sign of chronic hepatic diseases [154, 159].

Hepatic damage caused by halothane, hydrazines (iproniazid, INH), carbon disulphide and α-methyldopa are probably also due to LPO [455]. The mechanism of halothane hepatotoxicity has been extensively studied because of its use as an anaesthetic and the risks involved. It needs metabolic activation to become hepatotoxic and its biotransformation is mediated by cytochrome P-450. Depending on the oxygen tension, it is metabolized either via oxidative degradation (high O_2 tension), when its stable, nontoxic end products are excreted in the urine, or via reduction at low O_2 tension, or under anaerobic conditions. In the second pathway a radical CF_3CHCl is produced, which can initiate LPO by hydrogen abstraction, or can bind covalently to microsomal membrane components including cytochrome P-450. The decrease in the cytochrome P-450 level observed after exposure to halothane in a hypoxic atmosphere is due to LPO and covalent binding. Hypoxia may occur during anaesthesia, which enhances the hepatotoxicity of halothane. According to epidemiological studies, halothane-induced liver damage is more frequent in obese patients, which may be due to a decreased oxygen supply of the hepatocytes in such individuals compared with those of normal body weight. Owing to hypoxia, the damaging reductive pathway dominates [89, 109, 455].

14.3 Wilson's Disease

Increased $O_2^-\cdot$ concentration in the cells and extracellular space develops in Wilson's disease because: (a) coeruloplasmin is an extracellular scavenger with weak SOD activity, and the low coeruloplasmin level found in this disease signals a decreased antioxidant protection of the extracellular space; (b) owing to its ferroxidase property, coeruloplasmin catalyses the $Fe(II) \rightarrow Fe(III)$ oxidation of iron - if this catalytic action is absent, autooxidation of Fe(II) produces $O_2^-\cdot$; (c) finally, since free copper is deposited in the tissue as a result of coeruloplasmin deficiency, Cu^{2+} is reduced to Cu^+ with the help of SH groups and Cu^+ undergoes autooxidation, during which $O_2^-\cdot$ is formed (Fig. 38).

The occurrence of cirrhosis and extrapyramidal damage in Wilson's disease is evidence for the possible free radical origin of cirrhosis and extrapyramidal syndromes. Haemolytic episodes observed in Wilson's disease may be attributed to copper entering the circulation [404].

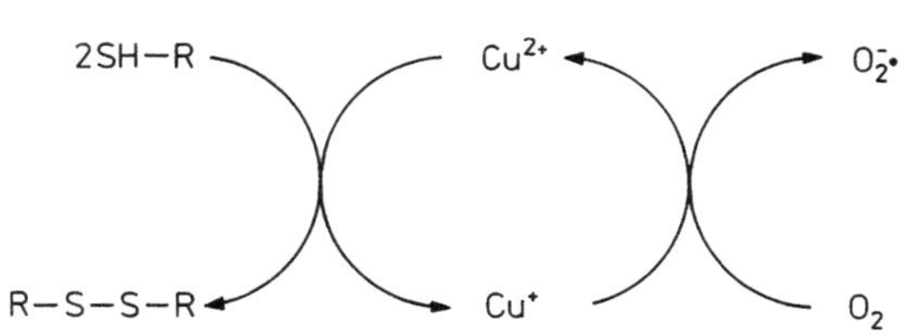

Fig. 38. One of the possible mechanisms of increased cellular and extracellular $O_2^-\cdot$ production in Wilson's disease: $O_2^-\cdot$ formation induced by copper deposited in the tissues. SH-R sulphhydryl; R-S-S-R disulphid. (After Nohl [404])

14.4 Haemochromatosis

It has been proved that the injuries caused by secondary haemosiderosis (transfusion, β-thalassaemia) are due to LPO. It is, therefore, probable that free radical reactions also play a role in the development of injuries of various organs in haemochromatosis. This is supported by lipofuscin and ceroid (indicators of in vivo LPO) having been found in patients with Wilson's disease, haemochromatosis and Dubin-Johnson syndrome. The favourable effect of D-penicillamine in Wilson's disease and haemochromatosis is due not only to its metal chelator effect, but probably also to the SOD activity of its copper complex [119, 256, 267, 404, 605].

14.5 Dubin-Johnson Syndrome

Dubin-Johnson syndrome is a hereditary form of conjugated hyperbilirubinaemia, characterized by the progressive accumulation of lipofuscin and melanin in hepatocytes. Selective deficiency of mitochondrial MnSOD was found in patients with Dubin-Johnson syndrome, while the cytosol activity (CuZn-SOD) was normal. The resultant increase in univalent leakage may be responsible for lipofuscin production and for the increased tumour frequency observed in this disease. Conjugated hyperbilirubinaemia is probably due to the damaging effect on the lysosomes of accumulated lipofuscin, because lysosomes play a role in the excretion of conjugated bilirubin. The reason for increased melanin production is not known, the catalysis of the $O_2^{-}\cdot$-dependent steps of melanin biosynthesis has been suggested. Morphological changes of the mitochondria indicate a mitochondrial origin [430].

14.6 Antioxidant Effect of Hepatoprotective Drugs

The therapy of chronic liver diseases has not been solved. Owing to the variety of causal factors in the pathogenesis of chronic liver lesions various drugs are being used. Some of them are applied in inflammatory reactions of the liver, others are used as cytoprotective agents in toxic liver cell alterations. However, there is no specific and unequivocally effective drug for these conditions.

In general we use drugs against the viral and bacterial infections, drugs for immune correction (for immune suppression or in other cases for immune stimulation), drugs for substitution (e.g. folic acid), as well as drugs for liver cell protection. The effectiveness of this treatment, however, can hardly be assessed either in human examinations or in animal experiments. In the following sections we discuss our results for various experimental models in the field of toxic liver lesions as well as in the therapy of chronic liver diseases.

14.6.1 Cyanidanol-3 and Lysosomal Enzyme Activity

It is known that in the serum of patients with liver diseases the activity of the mitochondrial and microsomal enzymes increase. This fact is used as a routine aid in clinical diagnosis. As for the diagnostic value of the lysosomal en-

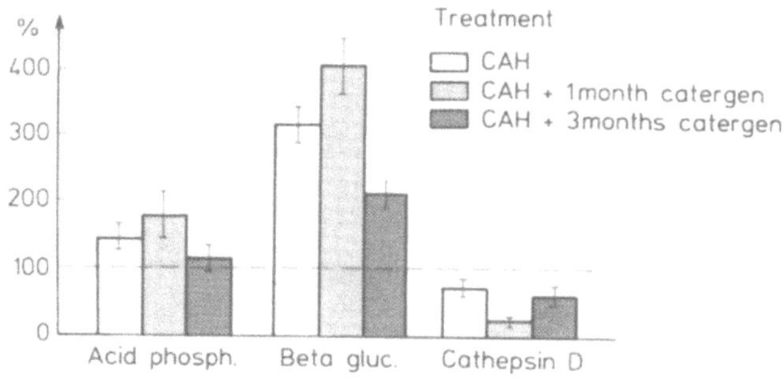

Fig. 39. Lyosomal enzymes in the serum of patients with chronic active hepatitis (CAH) treated with Catergen

zymes released from different cells into the serum, there are few useful data [42a]. We have demonstrated [151a] that the activity of the lysosomal enzyme in the serum was changed in chronic liver disease. The acid phosphatase and β-glucuronidase activity increased, while cathepsin D activity decreased. β-Glucuronidase enzyme activity of the granulocytes of patients with liver disease and the release of this enzyme from the granulocytes varied according to the different types of chronic liver disease, adequately reflecting the progression or regression of the disease.

In a clinical investigation we have studied lysosomal enzyme activity in the serum, the lysosomal enzyme content of the granulocytes and their release in chronic active hepatitis. In addition we have investigated the in vivo effect of Al-cyanidanol-3 (Catergen) treatment on the lysosomal enzyme activity, as well as its in vitro effect on granulocytes. Details of the methods have been published [157].

The results are shown in Figs. 39 and 40 and Table 2. In chronic active hepatitis the activity of acid phosphatase and β-glucuronidase in the serum significantly increased as compared with the controls. Cathepsin D activity did not change appreciably. At 1 month after Catergen treatment β-glucuronidase and acid phosphatase activity was consistently higher than in the controls. After treatment for a further 2 months acid phosphatase activity decreased to the normal level and the level of β-glucuronidase was also close to the activity found in healthy individuals, although it was still higher than normal. Cathepsin D activity was still significantly lower than that found in healthy persons (Fig. 39).

In patients with chronic active hepatitis (CAH) β-glucuronidase release from the granulocytes increased in vitro and the β-glucuronidase activity of the cells decreased. The rate of release was high compared with the controls (Fig. 40). After Catergen treatment for 1 month, β-glucuronidase enzyme release from the granulocytes decreased significantly, the enzyme content within the granulocytes increased considerably and the rate of release returned to normal values. During 3 months' treatment there was no appreciable change in these values, all three remaining within normal limits.

In the course of the in vitro examinations the behaviour of granulocytes

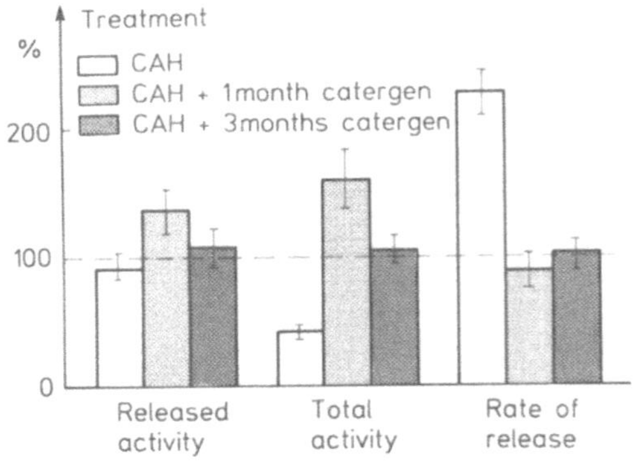

Fig. 40. β-Glucuronidase activity of granulocytes derived from patients treated with Catergen

derived from the blood of healthy individuals was studied in isotonic and hypotonic environments after incubation in Catergen. In an isotonic environment the release did not decrease significantly. In a hypotonic environment the lysosomal enzyme release increased considerably: in vitro incubation with Catergen in small doses (0.5 and 5.0 g/ml) decreased the degree of release significantly. In a higher dose (50.0 g/ml), Catergen proved to be toxic, since the lysosomal enzyme release increased to a degree similar to that observed in the hypotonic environment (Table 2). These results may be explained by the radical scavenging action of Catergen [505], which neutralizes the free radicals, exerts a protective effect on the lysosomal membranes and inhibits the generation of superoxide free radicals from the granulocytes.

In several cases cyanidanol-3 caused a number of adverse reactions. Among them were reactions involving the immune system, fever, haemolytic anaemia, skin reactions, gastrointestinal, and haemopoietic lesions and other miscellaneous alterations. In some cases the reaction was fatal and in 1985 cyanidanol-3 was withdrawn from the drugs approved for treatment of hepatic diseases.

Table 2. The effect of Catergen on the β-glucuronidase release of granulocytes in vitro

	Release rate[a]	Release related to control (%)
Control	2.31	100
Distilled water	8.80	381
Catergen 50 μg/ml	10.80	468
Catergen 5.0 μg/ml	2.00	87
Catergen 0.5 μg/ml	2.00	80
Distilled water + Catergen 50 μg/ml	9.40	407
Distilled water + Catergen 5.0 μg/ml	3.55	154
Distilled water + Catergen 0.5 μg/ml	2.90	126

[a] (released activity) (total activity) × 100

14.6.2 Dihydroquinoline-type Antioxidants and Acute Liver Lesions

Historically, recognition of carbon tetrachloride- and galactosamine-induced liver injuries has been based on morphological assessment with the light microscope. Recently, a number of enzyme abnormalities have been described and the role of free radical reactions has been suggested. Thus, simultaneous characterization with a biochemical marker of free radical reactions, seemed to provide a more refined approach to the understanding of these pathological processes.

The pharmacological action and possible therapeutic uses of some recently developed synthetic, nontoxic, dihydroquinoline-type antioxidants were studied. The effect of the lipid-soluble 6,6-methylene-bis-(2,2,4 trimethyl-1,2-dihydroquinoline) ($n = 1$, 2 or 3) (MTDQ) on carbon tetrachloride-induced acute liver injuries was investigated, and that of the water-soluble 5,6-methylene-bis-(2,2-dimethyl-4-methanesulphonic acid sodium 1,2-dihydroquinoline) (MDS) on galactosamine-induced acute liver injuries was investigated in CFLP mice (Lati, Hungary). Details have been published elsewhere [150].

Both the effect of CCl_4 and galactosamine enzyme induction of electron transfer enzyme systems can be observed, including the generation of superoxide radical anion $O_2^{-}\cdot$ and OH. According to present knowledge on free radical reactions in biological systems it is the OH which interferes with the integrity of biomolecules. Lipid-soluble antioxidants such as MTDQ, and water-soluble antioxidants such as MDS, inactivate free radicals generated by autooxidation and decompose endoperoxides.

After carbon tetrachloride treatment serum glutamic oxaloacetic transaminase (GOT) activity increased significantly ($P < 0.001$). With MTDQ and CCl_4 combined, GOT activity was found to be higher than that of untreated controls, but significantly lower than that observed in the CCl_4-treated mice ($P < 0.001$; Fig. 41).

Histopathological investigation of the livers of CCl_4-treated animals revealed, in most animals, the following characteristic pathomorphological changes: fatty degeneration, balloon cell degeneration, centrilobular fatty degeneration or necrosis, and vacuolization. The simultaneous administration

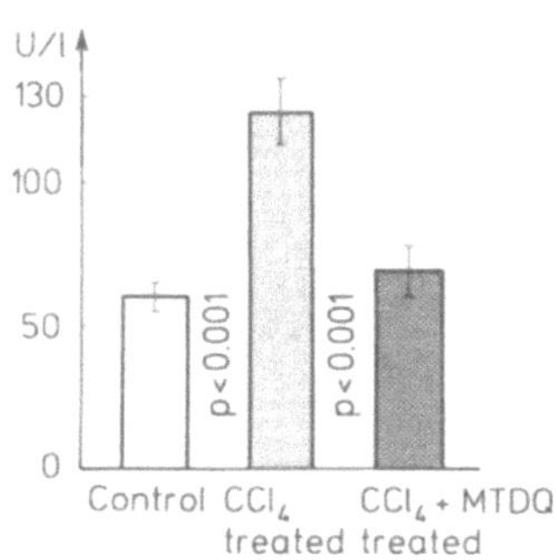

Fig. 41. The effect of MTDQ on serum GOT activity in acute CCl_4-induced liver injury

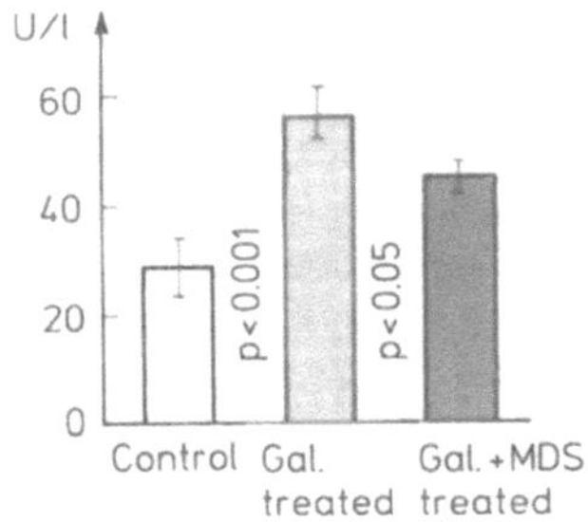

Fig. 42. The effect of MDS on serum GOT activity in acute galactosamine-induced liver injury

of MTDQ significantly diminished the morphological abnormalities in CCl_4-induced liver injury.

Galactosamine treatment caused a twofold increase of GOT activity compared with that measured in untreated controls ($P < 0.001$). With simultaneous galactosamine and MDS administration GOT activity was found to be higher than that of untreated controls, but significantly lower than that measured in the galactosamine-treated group ($P < 0.05$; Fig. 42).

The malondialdehyde concentration in liver tissue was significantly increased compared with untreated controls ($P < 0.01$). With simultaneous CCl_4 and MTDQ, as well as with simultaneous galactosamine and MDS, it decreased significantly ($P < 0.01$ and $P < 0.05$, respectively; Table 3).

With both CCl_4 and galactosamine, pathomorphological changes and biochemical alterations characteristic of CCl_4- and galactosamine-induced liver injuries developed. In the majority of CCl_4-treated mice fatty degeneration, balloon cell degeneration, vacuolization, centrilobular fatty degeneration and/or necrosis were observed. Galactosamine, however, gave rise to signs resembling those observed in viral hepatitis and other inflammatory liver diseases. Of the biochemical parameters, serum GOT activity and malondialdehyde concentration were chosen for this experiment. Both were found to rise with the two pathogenic substances. The simultaneous decrease and parallel changes of patterns of activity, malondialdehyde concentrations and pathomorphological liver lesions observed after MTDQ + CCl_4 and MDS + galactosamine treatment point to the preventive action of MTDQ and MDS.

Table 3. Malondialdehyde concentrations of the liver measured after administration of CCl_4, CCl_4 + MTDQ, galactosamine and galactosamine + MDS, compared with control values

Substance tested	Malondialdehyde (μg/g)
CCl_4	1.17
CCl_4 + MTDQ	0.65
Galactosamine	0.52
Galactosamine + MDS	0.31
Untreated control	0.14

14.6.3 Effect of AICA Phosphate on Lymphocyte Populations in Chronic Active Hepatitis

In the different forms of chronic active hepatitis therapy depends on the causative factors. Since immune deficiency is observed in some cases, immune stimulant therapy has been recommended. Levamisole treatment is widely used for this reason [424a]. We have shown that 4-amino-5-imidazolecarboxamide (AICA) had a beneficial effect on humoral immune reactions in chronic active hepatitis, decreasing the elevated concentrations of IgG [151a]. In addition we studied the effects of 4 weeks' therapy with 600 mg daily of the imidazole derivative AICA phosphate (AICA-P) on different cellular immune functions of patients with chronic alcoholic liver disease. We have found that in vivo AICA-P therapy does not modulate ADCC and natural killer (NK) activity. It inhibited lymphoblast transformation in patients with originally normal values of thymidine ^3H incorporation and enhanced lymphoblast transformation in patients with originally low responses. This phenomenon can be explained by the Helper T-lymphocyte/suppressor T-lymphocyte ($OKT4^+/OKT8^+$) rate-modulating effect of in vivo AICA-P therapy [107a].

14.6.4 Silymarin as a Free Radical Scavenger

Silymarin is an active flavolignan which owing to its free radical scavenger activity is a membrane stabilizing agent and an effective antioxidant (Fig. 43) [37, 68, 559a, 559b, 559c]. Silymarin is an isomeric mixture of silibinin, silidianin and silichristin (Fig. 44), silibinin being the most active component.

The therapeutic application of silymarin (Legalon) for treatment of patients with liver disease remains to be etablished. Recently a water-soluble form of silibinin (the dihemisuccinate sodium salt) has been developed [361 a, 559 b].

Silymarin and silibinin hemisuccinate sodium salt proved to be effective in animal experiments in the prevention of carbon tetrachloride, galactosamine and ethanol hepatotoxicity as well as *Amanita phalloides* intoxication [559a, 559b]. Their noncompetitive lipoxygenase and prostaglandin synthetase inhibiting activity and cellular immunoreactivity reducing effect is favourable in the therapy of chronic liver diseases. In clinical trials these compounds were effective in chronic hepatitis, cirrhosis, drug-induced liver toxicity (CCl_4, galactosamine, ethanol, tuberculostatics, *A. phalloides*, hypnotics, insecticides) [27, 98, 152a, 152b, 451].

The effect of silibinin, 1.0–10.0 g/ml on phytohaemagglutinin (PHA-) and concanavalin-A (Con-A)-induced lymphocyte blast transformation in normal individuals and patients with chronic alcoholic liver disease is shown in Figs. 45 and 46. With a suboptimal dose of PHA, silibinin exerted no significant effect on blastogenesis in a concentration range corresponding to the usual in vivo therapeutic concentration. However, with an optimal stimulatory dose of PHA, silibinin significantly decreased thymidine ^3H incorporation at a concentration of 10.0 g/ml. In the presence of 25 g/ml Con-A, silibinin inhibited in vitro lymphocyte blast transformation of the patients in a dose-dependent fashion.

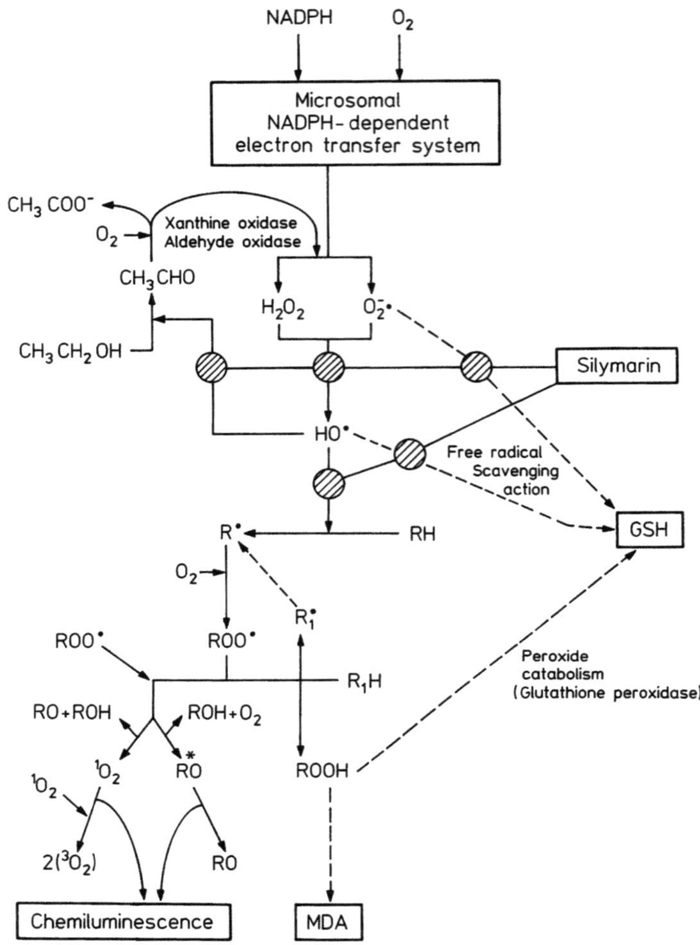

Fig. 43. Site of action of silymarin

Total superoxide dismutase (SOD) activities of the erythrocytes and lymphocytes of patients and of healthy individuals are summarized in Fig. 47. Patients with chronic alcoholic liver disease had significantly lower erythrocyte and lymphocyte SOD activities than the mean values of the healthy control group. After 1 month treatment with silibinin the originally decreased SOD values of the patients significantly increased both in erythrocytes and in lymphocytes.

Atherogenic diet significantly inhibited the Con-A-induced blast transformation of rat spleen cells in vivo. Silibinin reversed this reaction (Fig. 48). A similar effect of silibinin was found in lysosomal examinations of rat liver. The malondialdehyde (MDA) equivalent content of liver lysosomal membrane was increased in the animals fed atherogenic diet. The silibinin treat-

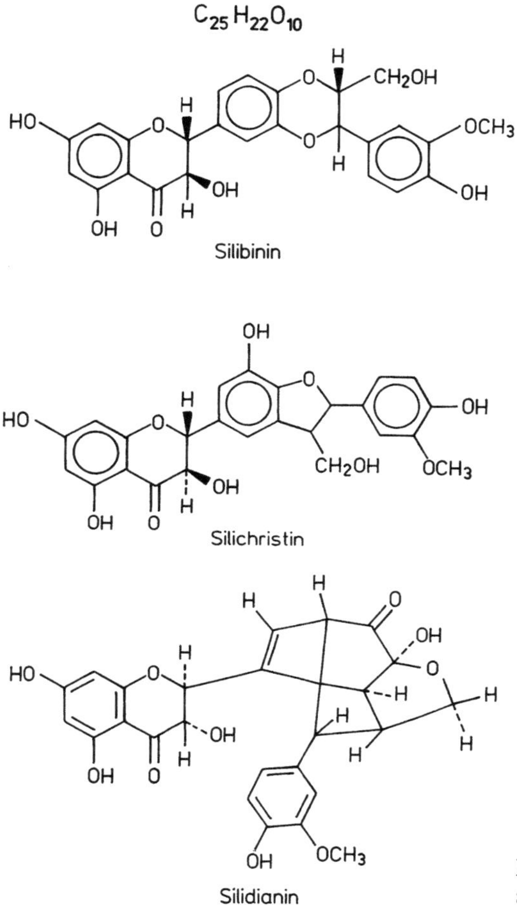

Fig. 44. Chemical structure of silymarin isomers

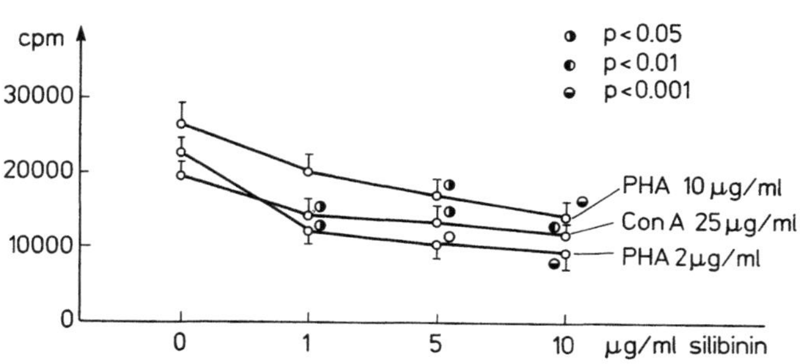

Fig. 45. Effect of in vitro treatment with silibinin on the lectin-induced lymphoblast transformation of healthy subjects

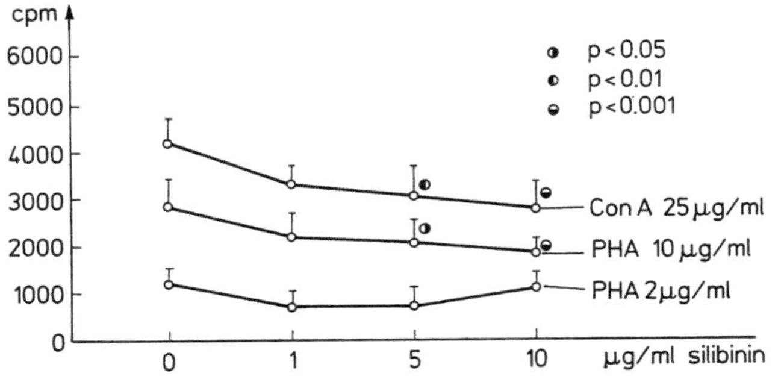

Fig. 46. Effect of in vitro treatment with silibinin on the lectin-induced lymphoblast transformation of patients with chronic alcoholic liver disease

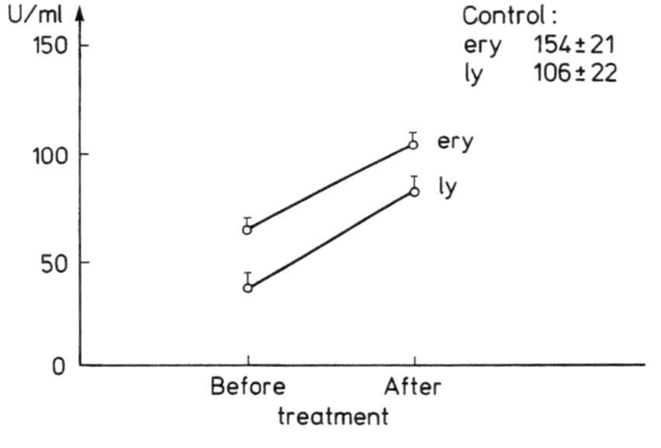

Fig. 47. Effect of treatment with silibinin on erythrocyte (ery) and lymphocyte (ly) SOD activity of patients with chronic alcoholic liver disease

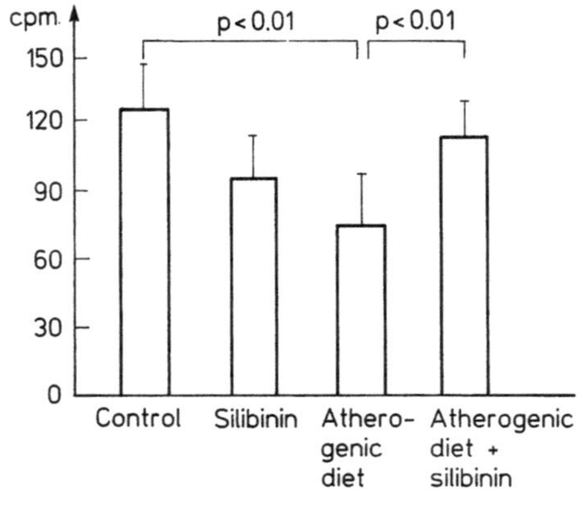

Fig. 48. In vivo effect of silibinin of the Con-A-induced (10 µg/ml) blast transformation of rat spleen cells

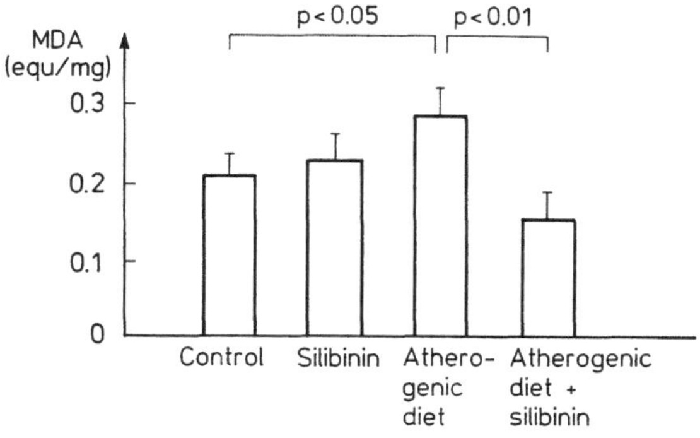

Fig. 49. Effect of silibinin on liver lysosomal fraction of rats on atherogenic diet

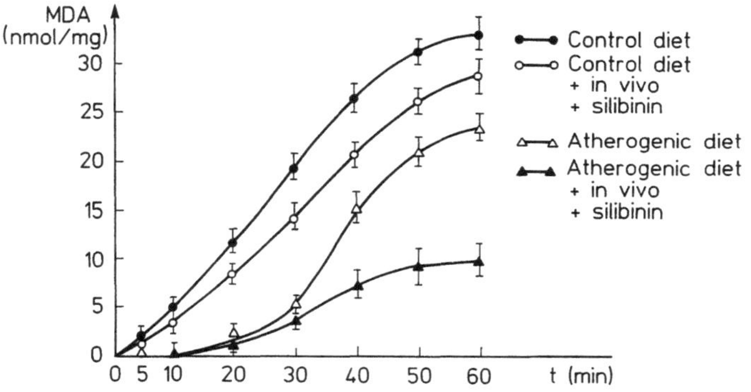

Fig. 50. In vivo inhibitory effect of silibinin on NADPH + iron-induced lipid peroxidation in rat microsomes (dose 25 mg/kg × 5 days)

ment normalized the MDA equivalent content of lysosomes (Fig. 49). Enzymatically induced lipid peroxidation was inhibited with in vivo silibinin treatment both in the control animals and in rats fed atherogenic diet (Fig. 50).

Silymarin and its water-soluble derivative probably exert their cytoprotective effect in two ways: (a) owing to their free radical scavenger activity they increase the tissue glutathione content and are potent antioxidants; and (b) by their interaction with cytochrome P-450 they block the biotransformation of drugs whose metabolites are cytotoxic [361a].

14.7 Summary

Detoxification of xenobiotic and toxic substances is an important hepatic function, in the course of which large amounts of ROIs are produced. This is why the liver is well supplied with antioxidants. The pathogenic role of LPO in alcohol-induced damage has been proved. This is also true for alcoholic liver disease. However, the mechanism whereby alcohol causes LPO is not known. It has been suggested by Lewis and Paton that xanthine oxidase plays a role in this. The effect of ethanol increasing microsomal enzyme activity is another possible mechanism. The third alternative is that acetaldehyde binds the SH-containing GSH precursors, thereby decreasing the hepatic GSH level. Haematological complications of alcoholic liver disease (macrocytosis, haemolytic anaemia, production of target cells and spur cells) may, at least in part, be due to free radical reaction. The role of LPO-induced damage in the pathogenesis of chronic liver diseases has been suggested, based on enzymatic studies (lysosomal enzymes in the serum). Carbon tetrachloride-induced liver damage, which is also due to free radicals, may be regarded as a model of hepatic injuries caused by xenobiotic agents. Pathological free radical reactions probably play a role in the hepatotoxicity of halothane, hydrazines (iproniazid, INH), carbon disulphide and α-methyldopa. The involvement of free radical events in the toxic effects of acetaminophen is postulated. Free radical reactions participate secondarily in the development of organ damage in Wilson's disease, haemochromatosis and secondary haemosiderosis. Sound experimental evidence is available to prove this. Selective deficiency of mitochondrial MnSOD and a consequent increase in univalent leakage has been suggested to be the primary factor in the pathogenesis of Dubin-Johnson syndrome, but only a small number of cases has been reported, in which such evidence could be obtained. The majority of drugs used in the treatment of liver diseases are antioxidants (vitamin E, lipoic acid, methionine, Legalon).

15 Free Radical Reactions in Diseases of the Gastrointestinal Tract

15.1 Ischaemic Intestinal Disease

Ischaemic intestinal disease is discussed in Sect. 8.

15.2 Necrotizing Enterocolitis

The morphology of early mucosal damage occurring in the necrotizing enterocolitis of neonates closely resembles that of free radical induced changes characterizing reperfusion ischaemia. Necrotizing enterocolitis is a severe disease of neonates, which occurs in cases of immaturity and stress. The stressor effects may include respiratory insufficiency, cold, shock, surgery and anaesthesia. The disease is characterized by extensive haemorrhagic necrosis of the small intestinal mucosa. It may progress, causing transmural necrosis and perforation. In anaesthetized newborn pigs the perfusion of intestinal mucosa decreased selectively in response to hypoxia and hypothermia. A similar picture was obtained by fluorescein fluorescence in a human neonate operated on for necrotizing enterocolitis. The histological lesion of the superficial small intestinal mucosa of anaesthetized puppies exposed to hypoxia was similar to that seen in adult dogs in the early phase of intestinal ischaemia. In puppies fed cow's milk the early histological lesion of necrotizing enterocolitis developed, as opposed to those given mother's milk. This may have been due to the higher xanthine oxidase level of cow's milk, which is a more likely explanation than the earlier theories based on differences in immunoglobulin levels

[420]. Dalsing's intestinal ischaemia in rats (see Sect. 8) is an excellent model of clinical necrotizing enterocolitis. The protective effect of SOD in this model is indirect evidence of the participation of free radical reactions in the pathogenesis of necrotizing enterocolitis.

15.3 Inflammatory Intestinal Diseases

Only sporadic data are available on the role of free radical reactions in inflammatory intestinal diseases (Crohn's disease, ulcerative colitis). Leukocytic infiltration of the intestinal wall is a common characteristic of these diseases, and the ROIs thus produced cause changes in mucosal permeability [136, 420].

15.4 Radiation Injury of the Intestines

Radiation injuries occurring during radiotherapy of malignant lesions are predominantly due to free radical reactions. HO· scavengers, and to a lesser extent SOD, effectively reduce the damage caused by ionizing radiation.

15.5 Gastric and Duodenal Ulcer

Only preliminary data are available on the part played by free radical reactions in the development of gastric and duodenal ulcer. Significantly increased spontaneous and induced chemiluminescence was found in these conditions as well as in chronic gastritis. In ulcer the lipid peroxide level increases in the serum and in erythrocytes. The inhibitory effect of lipid peroxides on cell division and regeneration may retard the healing of ulcers. A group of ulcer patients were given antioxidants in addition to the usual drug therapy, and a more rapid healing of the ulcers was found in this group. Cimetidine, Gapona (composite medicament containing trantelinium bromide, benactyzine, and Halidor) Peritol (cyproheptadine), propranolol and vitamin U have antioxidant effects in addition to their specific actions, as indicated by their decreasing effect on the elevated serum lipid peroxide level of ulcer patients. The antioxidant effect of propranolol has also been described in cardiac infarction, and this may play a role in the increased postinfarction survival observed after chronic propranolol administration [530].Vitamin E deficiency caused ulcers in human volunteers, whereas vitamin E treatment had a favourable effect in gastric ulcer patients [146, 513]. In rats exposed to stress (food deprivation) vitamin E administration delayed or prevented the development of stress ulcers and had a favourable effect on ulcers that had already developed [290].

Vitamin A and β-carotene have been described to have beneficial effects in experimental gastric mucosal injury [26, 384]. Agents damaging the gastric mucosal barrier, e.g. ethanol, produce a picture resembling that seen in postischaemic reperfusion damage of the small intestine. Haemorrhagic erosive gastritis due to stress not only resembles mucosal ischaemia morphologically, but is, in fact, associated with mucosal ischaemia. DMSO, a potent ·OH scavenger, was found in rats to provide significant protection against stress ulceration. Although DMSO is characterized by a number of various biological actions, its favourable effect in this case must have been due to its ·OH scavenger activity [82]. Opposite results were also reported: the synthetic antioxidants ionol and dibunol failed to influence gastric mucosal lesions produced in rats by ligation of the pylorus, although they were able to inhibit the increased LPO of gastric tissue [572].

15.6 Acute Pancreatitis

Following secretin stimulation Braganza and co-workers found a significant increase in the LPO products excreted in the bile phospholipid fraction and duodenal juice fatty acid fraction by patients with pancreatic disease, as compared with control patients suffering from other abdominal diseases. Practically no LPO products were found in the pancreatic juice. The authors offered the following explanation for these findings. The increased LPO products found after secretin stimulation in the bile of patients suffering from pancreatic disease were probably due to the activity of hepatic detoxifying enzyme systems (e.g. cytochrome P-450), which also serve as a source for the small amounts of peroxidized lipids excreted in the bile normally. Secretin is probably not only a stimulator of pancreatic juice secretion, but is also an important regulator of the transfer into the bile canaliculi of the metabolites produced in the smooth endoplasmic reticulum of hepatocytes, and increases the activity of microsomal enzymes. Thus, although pancreatic diseases may not be associated with hepatopathy, ROIs released in the course of increased hepatic microsomal enzyme activity may play a role in the development of diseases of the pancreas.

Pancreatitis often occurs following food and drink abuse causing secretin stimulation. Secretin stimulation used in the experiment mimicked this mechanism. Ethanol consumption, which is one of the most frequent causes of pancreatitis, can damage the liver via a free radical mechanism, and can also increase the activity of microsomal enzymes and thereby the amount of ROIs produced. The authors maintain that the effect of ROIs increases the concentration of LPO products excreted in the bile, causing pancreatic disease [48]. In hyperlipidaemia, another frequent cause of pancreatitis, steatosis hepatis may cause disturbances in hepatic function and LPO is also increased in the serum. These changes may contribute to the development of acute pancreatitis. The hyperlipidaemia-inducing effect of ethanol may be involved in its causative role in pancreatitis.

Sanfey and co-workers used ex vivo perfused dogs as a pancreas model. Hyperlipidaemia was modelled by oleic acid infusion, migrating bile stones and food abuse by partial obstruction of the bile duct and by secretin stimulation, and ischaemia (e.g. shock-associated ischaemia) by inducing 2-h ischaemia. They also showed in earlier experiments that capillary damage, indicated by increased capillary permeability, is a common trait of all kinds of experimental pancreatitis. The pancreatitis produced was characterized by oedema of the pancreas and by hyperamylasaemia. In view of the proved role of free radical reactions in intestinal ischaemia, it was attempted to influence experimental pancreatitis with scavengers. The pathological changes failed to develop in all these types of experimental pancreatitis in response to jointly administered SOD and CAT [420, 480].

The serum vitamin E level of pancreatitis patients was found to be significantly lower than in the controls. Diminished PUFA causes proneness to pancreatitis, especially to pancreatitis due to cholelithiasis. These data show that LPO increases in pancreatitis and that it increases the susceptibility to pancreatitis. The favourable effect of vitamin E in chronic pancreatitis has been reported [534].

15.7 Summary

Only sporadic experimental data are available to prove the role of free radicals in the aetiology of the diseases described in this chapter. These include the necrotizing enterocolitis of neonates chronic inflammatory intestinal diseases (Crohn's disease, ulcerative colitis), gastric and duodenal ulcer and acute pancreatitis. The role of free radicals has been proved in radiation injury of the intestines. The connection between free radicals and acute pancreatitis and gastric ulcers is a new and interesting field.

16 Free Radical Pathogenesis of Amyloidosis

Free radicals probably play a role in secondary amyloidosis, as indicated by the increased frequency of the disease with age and its association with diseases in the development of which free radicals are involved (RA, collagenosis, chronic inflammations). Amyloid production in the blood and vascular wall, where free radical reactions occur at measurable levels, is another link between amyloidosis and free radicals. Amyloid fibri, the chief constituent of amyloid, is composed of two kinds of protein. AL protein consists of fragments of the light chain of immunoglobulins, or probably of complete light chains, while AA protein is the product of the oxidative degeneration of connective tissue components. Because AA is also produced by fibroblasts, neutrophils and hepatocytes, it may originate from the microfilaments of the cells. The proportion of the two protein components of amyloid varies. In primary amyloidosis or in amyloidosis associated with multiple myeloma the AL protein dominates, while in secondary amyloidosis, the AA protein dominates. The role of free radical reactions in the pathogenesis of amyloidosis is indicated by the finding that the survival of LAF_1 mice with amyloidosis increased if they were fed an antioxidant diet.

Amyloid is normally deposited in the ageing organism, which might be due to the common pathogenesis of the two processes. Senile amyloidosis includes a number of states which show some general characteristics also found in other types of amyloidosis, but are peculiar in respect of the tissue localization of the depositions and of the protein components. Senile amyloidosis most frequently affects the heart, but depositions may also be found in the brain (e.g. the characteristic cortical plaques of Alzheimer's disease) and in the pancreas.

Antioxidants may be used with success in secondary amyloidosis, owing to their increasing effect on both humoral and cell-mediated immunity and to their enhancing effect on suppressor T-cell function. The favourable effect of DMSO has been reported [83, 229, 237, 238, 608].

16.1 Summary

Only sporadic data are available in respect of the free radical pathogenesis of amyloidosis. The favourable effect of the antioxidant DMSO has been described.

17 Diabetes Mellitus and Pathological Free Radical Reactions

CuZnSOD is present in higher concentrations both in human and in animal beta cells than in other cells of the pancreas. This suggests that the enzyme plays an important part in the homeostasis of beta cells [187, 188].

The *N*-nitroso diabetogenic drugs alloxan and streptozotocin, and the rodenticide Vacor probably damage the beta cells via $O_2^-\cdot$ production. Surplus $O_2^-\cdot$ also partially damages SOD, its own dismutating enzyme after some time (only CuZnSOD is damaged), which results in a decrease of the self-protection of beta cells against $O_2^-\cdot$. The protective effect of SOD, suppressing the effect of diabetogenic drugs in animal experiments also proves this mechanism [95, 163, 187, 188, 211]. Gandy and co-workers demonstrated the protective effect of an SOD-like copper complex in streptozocin-induced diabetes of rats [186]. Alloxan-induced free radical generation was demonstrated directly - with luminol-stimulated chemiluminescence - in the islet cells of rats, whereas chemiluminescence was not noted even after a longer period following streptozotocin administration [11]. The alloxan-induced increase in chemiluminescence was significantly stronger in the rat pancreatic islets than in the erythrocytes or hepatocytes [11]. Urate, which is a strong ·OH and RO· scavenger, is a potent cofactor of cyclooxygenase. Alloxan may be produced in the human body during the oxidation of urate. It had been supposed earlier that urate is not catabolized in humans, but in recent studies alloxan has been found in human tissues. In humans 20% of the labelled urate administered was found to be catabolized. The myeloperoxidase of neutrophils and cyclooxygenase may be responsible for this catabolism. Thus considerable amounts of alloxan may be produced in the human organism in hyperuricaemia and/or in the case of increased activity of catabolic enzymes. This may explain the frequent coincidence of hyperuricaemia and diabetes [108].

SOD may play a role not only in experimental diabetes, but also in the pathogenesis of human diabetes. In insulin-dependent diabetes (IDDM) some environmental noxa may play the initiator role, in addition to genetic predisposition. The environmental agent may be a β-cytotropic virus (Coxsackie B, rubeola virus), or a toxic agent inducing autoimmunity by causing antigen perturbation. In this, a role may be played by decreased suppressor T-cell activity, by cytotoxic antibodies and by cytotoxic T- or NK- (natural killer) cells, which produce diabetes by damaging the beta cells [59]. At what point may free radicals enter this mechanism?

It has been shown in both human and animal cell lines that SOD is needed for the establishment of an interferon-mediated antiviral state [440] and for the activity of suppressor T-cells [236]. Round cell invasion was demonstrated in the pancreas in IDDM. It consisted mainly of small lymphocytes and of some granulocytes and affected the beta cells in the first place. With the help of the chemotactic factors released owing to insulitis, the cytotoxic antibodies covering the beta cells activate the respiratory burst of neutrophils attracted to the site. The ROIs produced can partially inhibit SOD activity in the beta cells, and thus selectively damage these cells. The lymphokines of the T-cells present can also activate the respiratory burst [59, 187]. Some toxin may act as the other possible initiator. So far only the diabetic ketoacidosis-inducing effect of *N*-nitroso derivatives has been observed [321]. These substances are well-known carcinogens. They may be produced in the environment, in food, and also in the body from noncarcinogenic precursors under the effect of nitrosating agents (e.g. in pesticides, beer, dairy products, in meat and fish during frying and in the body under the effect of bacteria containing nitrate reductase, etc.).

So far the only evidence obtained as regards the role of toxins in IDDM is an epidemiological study made in Iceland, where the pathogenic role of smoked mutton, rich in *N*-nitroso components, has been suggested in the diabetes of chiefly male patients born in October, whose parents were believed to consume such traditional meals at Christmas oron New Year's Day. The action of various *N*-nitroso derivates resembles that of alloxan and streptozotocin [321]. The high serum lipid peroxide level found in diabetics [353], the decreased arterial PGI_2 production, which returned to normal in response to vitamin E in rats with streptozotocin-induced diabetes [294], the low vitamin E level and increased TXA_2 production of platelets of diabetics [293], the low SOD level found in the erythrocytes of diabetics [351], and the decreased CuZnSOD and MnSOD activity in the neutrophils of such patients [396], are all findings which prove the participation of free radical reactions in the pathogenesis of diabetes.

Free radical reactions may contribute to the development of complications of diabetes. Increased adhesion of erythrocytes to endothelial cells has been described in diabetics. The same change is also found in sickle cell anaemia in which free radical reactions play an important role. Increased red cell adhesion and selective PGI_2 inhibition may, at least in part, be responsible for the vascular complications of diabetes [570]. Also platelets producing the hyperactive TXA_2 and other active substances (e.g. growth factor) must play a role in the development of vascular complications [387]. Increased levels in the serum of the lysosomal enzymes β-glucuronidase and β-*N*-acetylglucosaminidase have been described in diabetics and in the chemically induced diabetes of rats [371]. On the other hand, Hägglöf and co-workers [220] investigated the enzymes of cellular protection (CuZnSOD, MnSOD, CAT,

GP) in the lymphocytes and erythrocytes of patients with IDDM, and failed to find characteristic changes.

The weak ·OH scavenger action of glucose was described in Chap. II Sect. 4. High blood glucose levels may play a role in the proneness to infections of incorrectly regulated diabetics, owing to a decreased efficiency of phagocytes [477]. Shah and co-workers [489] investigated the granulocyte function of diabetic patients. They found an increased $O_2^-\cdot$ production by nonstimulated lymphocytes as compared with controls. Nonstimulated chemiluminescence did not differ in the two groups, whereas both $O_2^-\cdot$ production and the increase in chemiluminescence were significantly lower in diabetics in response to stimulation with PMA and opsonized zymosan. The increased $O_2^-\cdot$ production of nonstimulated leukocytes in diabetic patients is supposed to be due to some factor in the serum of the patients, because the ROI production does not increase in response to the appropriate stimuli. This may play a part in the increased susceptibility to infections of diabetic patients. We do not know the clinical significance of the increased nonstimulated $O_2^-\cdot$ production and the nature of the serum factor responsible for it, nevertheless, they may be involved in the cardiovascular complications of diabetes, owing to their platelet aggregation inducing and vascular damaging actions.

A lipoxygenase pathway has been demonstrated in beta cells, in addition to the cyclooxygenase pathway. The cyclooxygenase cascade has been found to inhibit insulin release from beta cells and to stimulate the lypoxygenase cascade. Although endogenous arachidonic acid release and administration of arachidonic acid were found to stimulate both cascades, they increase insulin release. The same effects have been found for unsaturated fatty acids [365]. Arachidonic acid metabolism probably plays a role in the regulation of insulin release by glucose, by enhancing the glucose phosphatidylinositol cycle,[4] in which phosphoinositides are hydrolysed to phosphatidic acid, which can activate phospholipase A_2 and thus also the arachidonic acid metabolism either directly or via calcium mobilization. This leads to increased insulin release [33] PGE (a product of the cyclooxygenase pathway) has been reported to inhibit the acute insulin response to an increase in the glucose level [463]. The connection of free radical reactions with arachidonic acid metabolism has been proved, and thus they may also have a part in the pathogenesis of type II (non-insulin-dependent) diabetes. Pathological free radical reactions, the incidence of which increases in old age, may activate arachidonic acid metabolism by the mediation of phospholipase A_2, and thereby increase insulin release, leading to the depletion of the preformed insulin stores after some time.

---

[4] Phosphatidylinositol cycle: an increasing number of cells are known which respond to activation with an increased turnover of their phospholipids, especially of the inositol-containing lipids and of phosphatidic acid. This may be regarded as a general response of the cells to activation. The cycle has also been observed in Langerhans islet cells.

17.1 Summary

It has been proved experimentally that the N-nitroso diabetogenic drugs alloxan and streptozotocin and the rodenticide Vacor damage the beta cells by $O_2^-\cdot$ production. Alloxan may not only be an exogenous agent, but may also be produced during the oxidation of urat mediated by cyclooxygenase or myeloperoxidase. Free radical reactions may play a role in the development of human insulin-dependent and non-insulin-dependent diabetes, in its symptoms and complications. This, however, is still a hypothesis in spite of the pertinent experimental data.

18 Haematological Aspects

Erythrocytes are especially sensitive to oxidative damage, because they are exposed to high O_2 concentrations, their membrane is rich in PUFAs, and because they cannot resynthesize their damaged components. Therefore, the metabolic activity of erythrocytes is largely directed to defence against oxidative stress. In this defence, in addition to the systems described so far, erythrocytes also rely on haemoglobin, which completes the cytosol protection of the cells against ROIs.

In oxyhaemoglobin, polarization occurs owing to the transfer of an electron from the heme iron to the bound oxygen, whereby the Fe(II)-oxy state is replaced by a Fe(III)-superoxy state (Fig. 51). Under normal conditions, the transferred electron returns to the iron if the oxygen is released by the haemoglobin, and thus iron persists in its Fe(II) state. The electron returns to the iron only if there is no anion or water in the heme-containing part which could interfere with the transfer. Although the heme-containing structure is hydrophobic, random fluctuations of the surrounding globin part occasionally facilitate the entry of water into the heme-containing part. The entry of water, or of small anions causes $O_2^-\cdot$ release, while the heme iron loses an electron and is converted to Fe(III), producing methaemoglobin. Thus small amounts of ROIs are continuously produced in the cell even under normal conditions, resulting in the daily formation of methaemoglobin from 3% of the total haemoglobin. Methaemoglobin is reduced to haemoglobin by the enzyme methaemoglobin reductase, using NADPH (see Fig. 32).

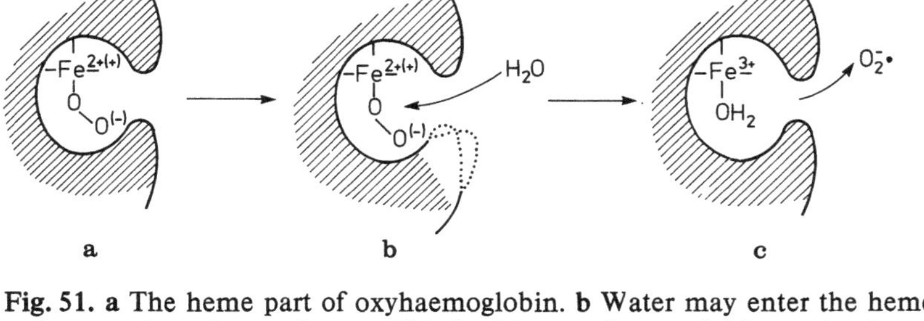

Fig. 51. **a** The heme part of oxyhaemoglobin. **b** Water may enter the heme part owing to minor fluctuations. **c** This leads to methaemoglobin production and $O_2^-\cdot$ release. (After Carrell et al. [66])

Any factor which causes increased flexibility or distortion of the heme-containing part of haemoglobin induces increased methaemoglobin production and $O_2^{-}\cdot$ release. Such factors include increased temperature, redox stains (ascorbate, acetylphenylhydrazine) and unstable haemoglobins. Thalassaemias also belong here because the isolated free chains of haemoglobin correspond to a special case of haemoglobin instability. Copper and iron enhance ROI production via their interaction with oxyhaemoglobin, considerably increasing the autooxidation of haemoglobin and $O_2^{-}\cdot$ production. There are also other sources of $O_2^{-}\cdot$ in the erythrocyte, e.g. free prophyrins, which form singlet oxygen radicals when exposed to UV light. Although haemoglobin is a source of $O_2^{-}\cdot$ under both normal and pathological conditions, it nevertheless constitutes a defence system in the cytosol of erythrocytes, which, beside the GP system, provides protection against oxidative stress.

Haemoglobin is exposed to oxidation at two sites. One is the heme part, where methaemoglobin is formed by reaction with H_2O_2. This reaction is limited to some extent by methaemoglobin reductase, which converts some of the methaemoglobin to haemoglobin, but if oxidant excess causes the destruction of the porphyrin ring, this is irreversible. The cysteinyl side chains of the globin molecule are the other target of oxidative attack. As a result of their oxidation, disulphide bridges are formed among the globin SH groups, leading to dissociation of the molecule to free subunits and to the denaturation of haemoglobin. This latter reaction is, fortunately, less likely because the SH groups of haemoglobin are less reactive than GSH, and their protection lasts until the depletion of GSH. Haemoglobin is thus protected against oxidation

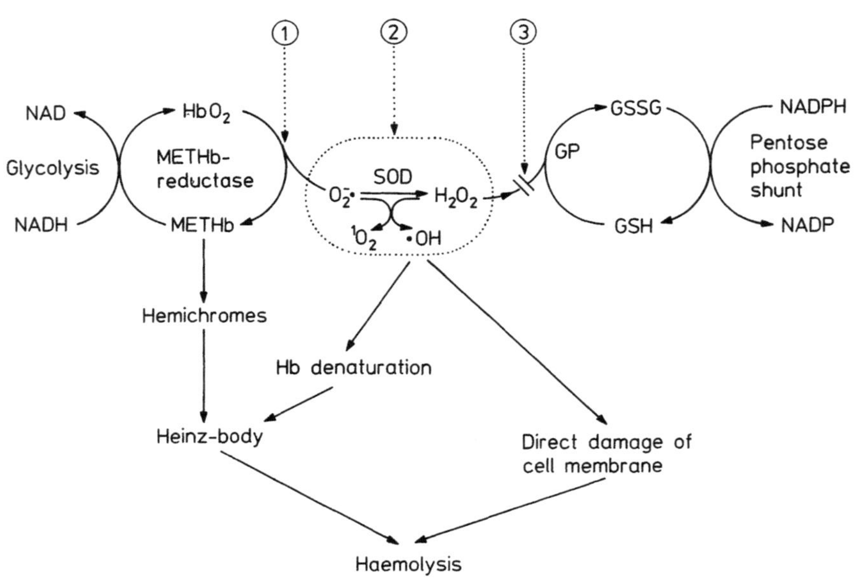

Fig. 52. Possible mechanisms of ROI formation under the influence of haemolytic factors and the mechanism of the resultant oxidative haemolysis, for explanation see text. (After Carrell et al. [66])

Table 4. Potential causes of oxidative haemolysis

Reactive oxygen intermediates	
Increased formation	Decreased protection
1. Via haemoglobin	3. Deficiencies of:
Unstable haemoglobins	Glucose-6-phosphate dehydrogenase
Thalassaemia	6-Phosphogluconate dehydrogenase
Oxidative drugs, e.g. Acetylphenylhydrazine; antimalarials,	Glutathione peroxidase
	Glutathione reductase
Divicine[a] (favism)	Glutathione synthetase
Heat	CAT
Metal ions	SOD
2. Direct	Vitamin E
Radiation	
Oxygen exposure	
Free porphyrin	

[a] Suggested mechanism

by its own quaternary structure and the reducing potential of the erythrocyte (GP, methyl reductase). In case of oxidant excess, haemoglobin is denatured, hemichromes (denatured ferrihaemoglobins) are produced, which are precipitated as Heinz bodies. ROIs are a hazard to the cell, directly by damaging the membrane and by the formation of Heinz bodies. Spontaneous hemichrome and Heinz body formation was observed, when distortion of the unstable haemoglobin molecule led to increased $O_2^{-}\cdot$ generation. The main factor responsible for the injury in this case was the formation of Heinz bodies. In thalassaemia the direct oxidative damage of the membrane is the major factor (Fig. 52, Table 4).

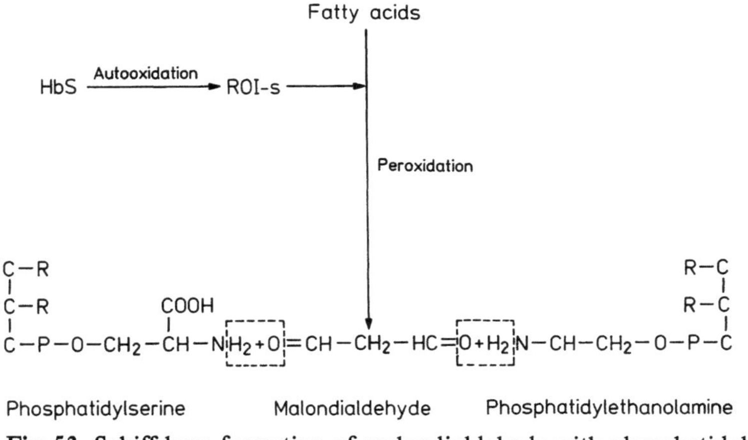

Fig. 53. Schiff base formation of malondialdehyde with phosphatidylserine and phosphatidylethanolamine. (After Shohet [495])

Heinz bodies cause haemolysis because the precipitated globin is bound to the red cell membrane by the disulphide bridges and causes a stiffness of the erythrocytes. Such cells are captured in the sinusoids of the spleen and are destroyed [66, 207, 586]. The biochemical mechanism of oxidative haemolysis can be illustrated with several examples taken from human pathology.

18.1. Unstable Haemoglobins

Sickle cell anaemia is the best known unstable haemoglobinopathy, one of a group of diseases characterized by changed amino acid structure due to point mutations at critical sites of the α- or β-chains. Sickle cell anaemia is characterized by increased spontaneous $O_2^- \cdot$ production in the sickle erythrocytes, decreased membrane vitamin E level, an increased amount of hemichromes, accelerated senescence of red cell membranes, fluorescent lipid pigments, elevated MDA level, decreased membrane PUFA content and by an increased copper level. As regards enzymatic defence, decreased CAT level together with increased SOD activity was found. Some authors described increased GP, while others found it decreased. Elevated SOD was probably induced by the increased spontaneous $O_2^- \cdot$ generation. Haemoglobin S has a stronger affinity to erythrocyte membrane, which means a higher risk of oxidant stress for sickle cells, because the soluble antioxidant system provides insufficient protection against the ROIs generated by the abnormal haemoglobin bound to the membrane, An ·OH production proportional to the hemichrome production was found in the sickle cells. Hemichrome had the role of the ion catalyst in the Fenton reaction [32, 78, 106, 245, 249]

In a series of elegant experiments Shohet and Jain demonstrated that in sickle cell anaemia MDA is incorporated in the erythrocyte membrane. MDA may enter into reaction with amino groups of the positively charged phospholipids of the membrane, forming a Schiff base (Fig. 53). MDA thus forms cross-links among the membrane phospholipids. Such a product, i.e. the Schiff base of MDA formed with phosphatidyl serine and phosphatidyl ethanolamine, was demonstrated in vivo by thin-layer chromatography of membrane lipids of sickled erythrocytes. When MDA was added in vitro to normal erythrocytes from controls and sickle cell anaemia patients, K leakage from the cells and a significant decrease in cell deformability could be observed. Both the disturbance in cation transport (K leakage) and the decreased deformability are well-documented defects of the membrane of the irreversible sickle cells (which do not regain their normal biconcave shape after reoxygenation). This does not mean that the Schiff base formed by MDA with membrane lipid is solely responsible for these membrane changes. MDA undoubtedly can enter into cross-linkage with the protein components of the membrane. High molecular weight protein aggregates were indeed demonstrated after MDA had been added to erythrocytes in vitro [299, 496]. Changes indicating the presence of such MDA Schiff bases were found by Hochstein and Jain in aged erythrocytes. These may play a role in the destruction of aged erythrocytes, which die in the spleen owing to their decreased deformability [260].

The data of Snyder and co-workers confirm the results of Hochstein and Jain as regards the role of oxidative processes in erythrocyte ageing: they demonstrated, besides the presence of irreversible spectrin cross-links, the binding of haemoglobin to the spectrin of ageing erythrocytes [509].

These changes play a decisive role in the pathogenesis of sickle cell anaemia. Increased $O_2^{-}\cdot$ production induced by the abnormal haemoglobin and the consequent damage of the antioxidant defence of sickle cells, i.e. increased LPO associated with a depleted defence system, probably contribute to the shorter life span of sickle cells (haemolysis). It has been observed that an increased number of deformed, shrunken erythrocytes may be found in vitamin E-deficient states (premature infants, cystic fibrosis, vitamin E-deficient adults). These changes could be suppressed by vitamin E administration.

Pathological free radical reactions contribute to the production of irreversible sickle cells. This is due, according to one theory, to the deformity of the structural proteins of the erythrocyte membrane, the spectrin-actin lattice. Another theory maintains that dehydration of the cells resulting from the abnormal ion permeability of the membrane leads to irreversible sickling. The two theories are not mutually exclusive, both changes may be due to peroxidative damage. The spectrin-actin lattice may be deformed by MDA formed as a result of LPO, or by other reactive aldehydes, which are able to produce cross-linkage between the free amino acid groups of proteins. Lattice deformity may also be due to the effect of membrane-bound haemoglobin S. Its effect on cation permeability has already been described. The participation of LPO in the production of irreversible sickle cells is directly supported by the in vitro observation that oxidants (e.g. H_2O_2) increase the production of these cells, while vitamin E partially inhibits it. The decreased deformability of sickle cells, which is due to their peroxidative damage, may contribute to their haemolysis, while it also plays an important role in the painful episodes of vaso-occlusive crisis and in the development of tissue infarctions caused by capillary obstruction [79].

The role of oxidant stress in the production of vaso-occlusive crisis is indicated by the first event of the crisis being the transformation of red blood cells to echinocytes. Part of the echinocytes were irreversible sickle cells, another part non-irreversible sickle cells and normal erythrocytes. There is a negative correlation between the proportion of echinocytes and the GSH content of the cells. The number of apparently normal erythrocytes increases after the crisis. The echinocyte-inducing effect of oxidant stress is proved by the finding that during density gradient ultracentrifugation the proportion of echinocytes was highest in the densest layer, which contained the irreversible sickle cells characterized by the most intensive ROI production. Also, echinocytic transformation could be induced by adding H_2O_2 to erythrocytes in vitro. Other authors found that the ability of erythrocytes to regain their normal shape (also from the echinocytic form) depends on the activity of the hexose monophosphate shunt, it decreases in oxidant excess. In the experiments cited, the number of irreversible sickle cells which did not undergo transformation to echinocytes did not change during the crisis [6, 567].

In rare unstable haemoglobinopathies (Hb Köln, Hb Sidney, Hb Christchurch) haemoglobin precipitation and increased Heinz body formation were induced in vitro by heating and incubation with activated phagocytes or erythrocytes. Haemolytic crises in patients suffering from these diseases occur via the same mechanism in hyperpyrexia and infections [585].

18.2 β-Thalassaemia

Increased LPO has been demonstrated in β-thalassaemia, and increased $O_2^-\cdot$ production was found in vitro. In β-thalassaemia, $O_2^-\cdot$ generated during the oxidation of excess haemoglobin α-chains, the iron released during the decomposition of the α-chains and the low haemoglobin content of the cells contribute to the increased $O_2^-\cdot$ production (Fig. 54).

A decreased PUFA content of the red cell membrane has been demonstrated in thalassaemia. After exogenous oxidant stress the number of SH groups decreased in the membrane, indicating cross-linkage. Decreased serum vitamin E level has also been described. Elevated SOD, CAT and G le-

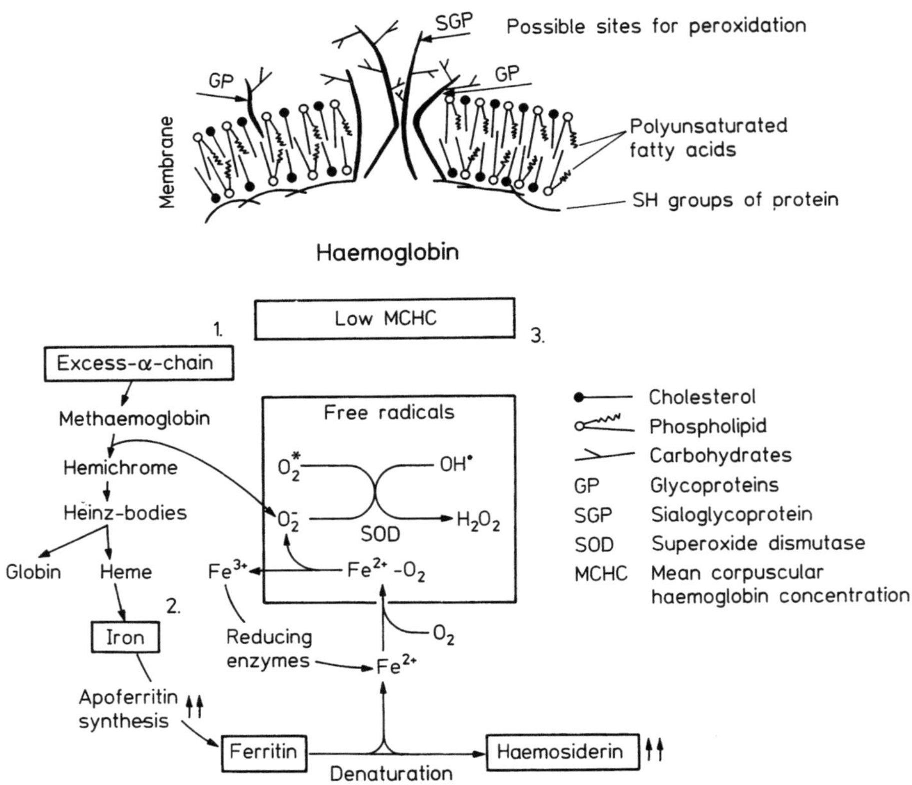

Fig. 54. Schematic representation of a thalassaemic erythrocyte showing the possible mechanisms of increased free radical production and the damaging effect of free radicals on the erythrocyte membrane. (After Rachmilewitz et al. [447])

vels were found in the minor form, probably owing to increased ROI production. In the major form, however, the level of these enzymes was normal, owing to the multiple transfusions the patients received. The number of transfusions could be reduced by parenteral vitamin E administration. Oral treatment was found ineffective [196, 199, 447].

18.3 Glucose-6-phosphate Dehydrogenase Deficiency

G-6-PD-deficient patients are known to be more sensitive to oxidative stress. Primaquine and pamaquin cause haemolysis in such patients [122, 197]. There is an interesting relation between unstable haemoglobinopathies (haemoglobin S, C, D, E), thalassaemia, G-6-PD deficiency and *Plasmodium falciparum* malaria: patients suffering from these diseases are immune to falciparum malaria. The increased intracellular release of ROIs must play a part in this immunity, since this is a common characteristic of these diseases (see Sect. 20).

18.4 Erythropoietic Protoprophyria

Dermatological manifestations (photosensitivity) and haemolysis in erythropoietic protoporphyria (EEP) are due to the production of an excited state of protoporphyrin by absorption of UV light. In the presence of molecular oxygen, this protoporphyrin releases energy, forming singlet oxygen or other ROIs. This is proved by the demonstration of LPO and inhibition of erythrocyte acetylcholinesterase in erythrocytes form EPP patients exposed to UV irradiation in vitro. Treatment of EPP with singlet oxygen scavenger and antioxidant vitamin E and β-carotene yielded favourable results [205, 424].

18.5 Paroxysmal Nocturnal Haemoglobinuria

In paroxysmal nocturnal haemoglobinuria (PNH) red blood cells show an increased sensitivity to haemolysis, which is probably due to some membrane defect. Decreased red cell acetylcholinesterase activity was found in PNH, which has no pathogenic role, but indicates a change in the membrane lipid environment caused by peroxidative damage. Contradictory data have been published as regards the presence of LPO and the usefulness of antioxidant therapy in PNH [417, 502].

18.6 Iron Deficiency Anaemia

Erythrocyte life span was investigated with ^{51}Cr and was found shorter in both humans and animals with iron deficiency anaemia. The erythrocyte shows increased sensitivity to peroxidative damage in iron deficiency because it is characterized by: (a) decreased GP, CAT and vitamin E levels: (b) increased protoporphyrin and iron levels: and (c) decreased haemoglobin content.

In addition to these changes, the presence of the MDA-phospholipid complex mentioned in connection with sickle cell anaemia and of high molecular weight protein aggregates indicating peroxidative damage were demonstrated in the red cell membrane. Membrane cation changes, also observed in sickle cells, were described, leading to a decrease of the intracellular K^+ content and an increase of the Na^+ content of erythrocytes. Decreased deformability caused by membrane rigidity - and not the unfavourable surface: volume ratio as believed earlier - was also found. Thus there are sufficient data available to suggest the causative role of peroxidative damage in the shortening of the life span of iron-deficient red cells. Nevertheless, further studies are needed for the clarification of the origin of LPO [69, 125, 276, 590].

18.7 Spur Cell Anaemia

LPO is probably involved in the development of spur cell anaemia occurring in severe hepatic disease (cirrhosis). It may also have a role in the pathogenesis of the primary disease and may be stimulated by vitamin E deficiency caused by decreased hepatic synthesis of factors necessary for the absorption of vitamin E (β-lipoprotein). Spur cells easily undergo haemolysis owing to changes in their membrane [495].

18.8 Abetalipoproteinaemia

In abetalipoproteinaemia, which is a rare hereditary disease, acanthocytes resembling the spur cells described in Sect. 18.7 are found. β-Lipoprotein is necessary for the plasma transport of vitamin E. The vitamin E level is decreased in these patients and its deficiency is probably responsible for acanthocytosis [386, 495].

18.9 Haemolytic Anaemia of the Newborn

The involvement of vitamin E deficiency has been proved in the haemolysis of neonates. Good results have been achieved with intramuscular vitamin E therapy [215].

18.10 Fanconi's Anaemia

Decreased leukocytic and erythrocytic SOD levels have been found in Fanconi's anaemia, which is associated with panmyelopathy in children and is characterized by a high frequency of chromosome aberrations and malformations. Based on the "spontaneous" chromosomal damages, a deficiency of DNA repair has been held responsible for the disease. The decreased SOD level, however, suggests an alternative explanation: the increased tumour frequency and chromosome aberrations might be due to increased cellular $O_2^{-}\cdot$ concentration. This possibility is supported by the observation that Fanconi's anaemia cells cultured in the presence of SOD and/or CAT showed fewer chromosome aberrations [283, 594]. Raj and Heddle found that SOD, CAT and L-cysteine added to FA cells reduced both spontaneous and mytomycin C-induced chromosomal breakage [448].

18.11 Summary

Erythrocytes are especially sensitive to oxidative damage, because they are exposed to high oxygen concentrations, their membrane is rich in PUFAs, and because they cannot resynthesize their damaged components. Oxyhaemoglobin has recently been found to be Fe(III)-superoxy state and not in the Fe(II)-oxy state. Under normal conditions, polarization ceases with the return of the shared electron to the iron, and iron persists in the ferrous state. Even under physiological conditions water may enter the heme-containing part of haemoglobin owing to random fluctuations of the globin part, although the heme-containing structure is hydrophobic. Water or small anions prevent the electrons from returning to the iron, leading to $O_2^-\cdot$ *release, while the heme iron loses an electron and is converted to·Fe(III). In this way small amounts of methaemoglobin are produced even under physiological conditions. Methaemoglobin is reconverted to haemoglobin by methaemoglobin reductase. Any factor causing distortion of the heme-containing part of haemoglobin leads to increased methaemoglobin production and* $O_2^-\cdot$ *release. This factor may be either high temperature, redox stains, unstable haemoglobins, iron, or copper. ROIs in the erythrocytes may also be produced by free prophyrins generating* $O_2^-\cdot$ *and other ROIs by absorption of UV light.*

Haemoglobin not only induces ROI production, but it also constitutes a defence system against oxidative stress in the red cell cytosol (besides the GP system) by protecting red cell integrity via autooxidation. In considerable antioxidant excess haemoglobin is denatured and is precipitated in the form of Heinz bodies. ROIs may damage the cell membrane directly or by forming Heinz bodies. The latter case is especially dangerous because the precipitated globin is bound to the cell membrane by disulphide bridges, causing rigidity of the erythrocytes. Such cells are captured in the sinusoids of the spleen and are destroyed.

The role of free radical reactions has been proved in haemolysis in sickle-cell anaemia, thalassaemia, erythropoietic protoporphyria, glucose-6-phosphate dehydrogenase deficiency, iron deficiency anaemia, abetalipoproteinaemia, and in the haemolytic anaemia of neonates. ROI formation and consequent oxidative haemolysis is due to the unstable haemoglobin in sickle cell anaemia; to a disturbance in haemoglobin production and the low haemoglobin content of the cells in β-thalassaemia; to decreased antioxidant defence in glucose-6-phosphate dehydrogenase deficiency; to the induction by UV light of excited protoporphyrin in erythropoietic protoporphyria; and to decreased antioxidant protection, increased copper and porphyrin levels and haemoglobin content of red cells in iron deficiency anaemia. In the haemolytic anaemia of the newborn, vitamin deficiency is responsible for the lower antioxidant protection. In paroxysmal nocturnal haemoglobinuria, spur cell anaemia and Fanconi's anaemia, the role of free radical reactions is only hypothetical.

19 Free Radical Reactions in Neonatal Diseases

Neonates, and especially premature infants, have a weak defence against oxidative stress: their defence systems are defective at almost all levels. The physiological barriers in the way of endogenous oxygen transport are immature.

Iron and copper, which catalyse LPO, accumulate in the fetal tissues in the last weeks of gestation. The levels of vitamin E, selenium, transferrin and coeruloplasmin decrease. Oxygen therapy, and also phototherapy used to treat hyperbilirubinaemia, may increase the risk of oxidative damage in neonates with defective defence, because of increased ROI production [310].

19.1 Neonatal Diseases Connected with Oxygen Toxicity

19.1.1 Haemolytic Disease of the Newborn

Haemolytic disease is discussed in Sect. 18.9

19.1.2 Retrolental Fibroplasia

The aetiological role of oxygen toxicity in retrolental fibroplasia (RLF) has been proved by several authors. It develops in two phases. The early changes develop in the least mature part of the retina, i.e. in the temporal region (the maturation of the retina proceeds centrifugally). Hyperoxia has a vaso-occlusive effect on the vasculature of the immature retina, producing retinal ischaemia characterized by arteriolar contraction and damage of the endothelium of the capillaries of the retina. Hyperoxia thus only induces early vasoconstriction, and RLF is paradoxically an anoxic retinopathy, although it is elicited by hyperoxia. The endothelium of the immature cells is sensitive to the ROIs generated first by hyperoxia, then by hypoxia, either because the activity of the protective enzymes is low, or because it has not been gradually exposed to oxygen, owing to the absence of maturation, during which exposure the activity of the protective enzymes could have increased. The second, vasoproliferative, phase is not caused directly by hyperoxia, but by the primary lesion. Progression of vasoproliferation leads to impairment of vision and to blindness.

With the advent of the intensive care of neonates the importance of this disease has grown. Hugarian authors described the use of D-penicillamine in premature infants needing oxygen therapy, which decreased the incidence of the disease. The SOD activity of the retina of newborn rats was found lower than other tissues (e.g. liver). Hyperoxic treatment induced only a small elevation of SOD activity in the retina. MTDQ, D-penicillamine and vitamin E were found to mitigate the severe complications of RLF [251, 258, 284, 292, 578]. Vitamin E is active in the hyperoxic phase (direct antioxidant effect) and it also inhibits vasoproliferation owing to its well-known inhibitory effect on collagen synthesis and wound healing [578].

19.1.3 Bronchopulmonary Dysplasia

In the aetiology of this progressive pulmonary fibrosis leading to severe respiratory insufficiency, oxygen toxicity and chronic mechanical respiration play an important role besides other factors. The pathological changes in the lung correspond to those described in Sect. 1, with some quantitative differences. Fibrosis is the final result of the early changes. Ehrenkranz and co-workers described favourable results of vitamin E treatment: it mitigated the progression of the disease, and prevented the development of severe lesions [131, 613]. In subsequent studies the same authors found vitamin E ineffective and attributed the discrepant results to differences in patient groups. The technique of artificial respiration had been improved meanwhile, and the newborns in the second series were born with a much less severe vitamin E deficiency, owing to the inclusion of vitamin E in the diet of the mothers. The conclusion was therefore modified, maintaining that vitamin E therapy is beneficial in this disease if there is vitamin E deficiency [130]. The severity of bronchopulmonary dysplasia was reduced in children suffering from respiratory distress syndrome by Rosenfeld and co-workers, who administered bovine SOD parenterally. They found no significant side effects [468].

19.1.4 Hyaline Membrane Disease

Hyaline membrane has been shown to be an LPO product. Experimentally it can be produced by ionizing radiation. It is known that oxygen radicals may damage type II pneumonocytes which have the function of producing surface active substances in the lung. LPO damage may thus be the primary cause of the disease. LPO may be due to the immaturity of the defence systems and to increased free radical production caused by asphyxia in utero or during labour. Hypoxia and the oxygen therapy applied to correct it may increase the damage. Therefore, it has been attempted to suppress free radical formation in hypoxia with Milurit (allopurinol), which inactivates xanthine oxidase [310].

19.1.5 Persistent Pulmonary Hypertension in Neonates

The syndrome is characterized by hypertension appearing a few hours after birth and persisting thereafter and by severe hypoxia. The presence of LTC_4, LDT_4 and LTE_4 (called collectively SRS-A) was demonstrated in the fluid used for rinsing the respiratory tract of neonates. The effect of these substances may account for the changes characterizing the disease. The neonates were given O_2 therapy after birth because of postnatal hypoxia. The appearance of SRS-A was probably due to stimulation of arachidonic acid metabolism by the increased production of free radicals, which, in turn, was caused by hypoxia and ventilation barotrauma [523].

19.2 Hyperbilirubinaemia of Neonates

Bilirubin is a known photodynamic agent producing $^1O_2$ in the presence of O_2 and light. Therefore, the phototherapy applied in hyperbilirubinaemias may be dangerous (mutagenic and carcinogenic effects, photohaemolysis, etc.). Jaundiced Gunn rats were irradiated with blue fluorescent light and the serum bilirubin level failed to decrease in some of the animals, although increased bilirubin excretion was demonstrated. Bilirubin produced as a result of haemolysis induced by the photosensitization of erythrocytes stabilized the serum bilirubin level in spite of increased excretion caused by phototherapy. This experiment and the well-documented in vitro photohaemolytic effect on erythrocytes of bilirubin suggest that photohaemolysis may occur in vivo as a result of the phototherapy of neonates [100].

The antioxidant D-penicillamine has been used with success in the treatment of hyperbilirubinaemia. Iron and other heavy metals which accumulate in the fetal tissue enhance, via enzyme induction, the microsomal heme-oxygenase activity, which is the rate-limiting step of heme degradation, but, at the same time, inhibit mitochondrial δ-aminolevulinic acid synthetase (ALAS) activity, which is the initial step of heme synthesis. Thus the balance is shifted towards breakdown and bilirubin accumulates. The release of iron in the fetal tissues is due to the haemolysis of fetal erythrocytes with increased liability to injury. The proneness of fetal erythrocytes to peroxidative haemolysis, in turn, is due to several factors, such as vitamin E deficiency and the increased copper content and dereased SOD and GP activity of fetal erythrocytes. The favourable effect of D-penicillamine (DPA) is due to the inhibition of bilirubin synthesis and stimulation of physiological conjugation of bilirubin. DPA inhibits bilirubin formation by: (a) increasing the osmotic resistance of fetal erythrocytes via its membrane stabilizing effect; and (b) decreasing the HO activity and the peroxidative damage of erythrocytes via its chelating action. In neonates the decrease in the activity of the uridine diphosphate glucuronyltransferase (UDPGT) system, which mediates the conjugation of bilirubin, is elicited by LPO induced by the iron and copper accumulated in fetal tissues. The chelating activity of DPA and the SOD-like effect of its copper complex can stimulate the enzyme by inhibiting LPO [311, 413].

19.3 Summary

The defence system against oxidative stress is deficient in neonates, especially in preterm infants. Oxygen therapy and the consequent ROI release play an aetiological role in severe diseases such as retrolental fibroplasia accompanied by severely impaired vision or blindness, and progressive pulmonary fibrosis characterized by severe respiratory dysplasia. The involvement of pathological free radical reactions has also been suggested in the pathogenesis of hyaline membrane disease and the persisting pulmonary hypertension of neonates, although sufficient proof is still lacking in this respect. In the hyperbilirubinaemia of neonates the increased susceptibility of neonatal erythrocytes to peroxidative haemolysis plays a role. This, in turn, is due to vitamin E deficiency, the increased copper content and decreased SOD and GP levels of the fetal erythrocytes. The favourable effect of D*-penicillamine in neonatal hyperbilirubinaemias has been reported.*

20 Free Radical Reactions in Infectious Diseases

It has recently been suggested that ROIs play a role in several phases of the interaction between infectious agents and the human body, and also in the effect of antimicrobial agents.

Some parasites (e.g. ascaris) and bacteria produce ROIs. An important role has been attributed to the balance between ROIs and the defence systems in the inactivation of infectious agents and in their virulence. ROIs produced by phagocytes (neutrophils, monocytes, macrophages, eosinophils) are involved in the killing of bacteria and parasites. It has been shown in a number of experiments that ROI production is the major antimicrobial mechanism of the macrophages when killing *Trypanosoma cruzi, Toxoplasma gondii* and *Leishmania* protozoa. The virulence of bacteria and parasites depends on the degree of the activation of the respiratory burst of the specific phagocytes (this, in turn, depends on whether the species is coated with specific antibodies or opsonin, which enhance activation) and also on the degree of antioxidant protection of the microorganism. If the degree of activation of the respiratory burst is high and the microorganism is well protected, its virulence is higher [120, 161, 194, 240, 397, 429, 434].

The mechanisms of action of several antiparasitic agents have been found to rely on ROI production, or they may also act by being converted to free radicals, e.g. the quinone and quinoline derivatives, the trypanocide lapachone, menadione and menoctone. The antracycline antitumour antibiotics with quinone and hydroquinone structures (daunomycin, doxorubicin, mitomycin C, actinomycin D), which are known to produce ROIs during their microsomal activation also possess trypanocide action. Gossypol, a phenol derivative, has a toxic effect on *T. cruzi* and *Plasmodium.* In China it is also used as a male contraceptive. Both its antiparasite and its spermicide actions are probably due to ROI production. The toxic effect of phenazine methosulphate on *T. cruzi* and *Leishmania* is probably due to a similar mechanism. The nitro derivatives nifurtimox, benznidazole and Klion (metronidazole) used with success in the therapy of Chagas' disease are first converted to $RNO_2\cdot$ radicals, which react with molecular oxygen, producing ROIs. The trypanocide effect of Bengal red and crystal violet is also due to ROI production. Chlorpromazine has been found to be effective against *Leishmania* and *Trypanosoma, brucei* after transformation to an active free radical cation [97, 120, 121, 379, 380, 392]. The favourable effect of haematological disorders (thalassaemia, sickle cell anaemia, unstable haemoglobinopathies - haemoglobin C, D, E - and G-6-PD deficiency) against malaria has been shown to be due to increased intracellular release of ROIs. *Plasmodium berghei* causing malaria in mice does not contain SOD and CAT, but engulfs by endocytosis the SOD of the erythrocytes of the host. *Plasmodium falciparum,* causing malaria in humans acts in a similar way with the difference that this organism has a low endogenous SOD activity. The importance of the oxidant-antioxidant balance in the defence against malaria is indicated by the in vivo antimalarial activity of various ROI-producing substances including alloxan, t-butylperoxide and pamaquin. The haemolytic effect of primaquine in G-6-PD-deficient patients is due to ROI formation. The dependence of *Plasmodium* on host red cell SOD suggests that these organisms might also be eliminated with the help of SOD inhibitors. The antimalarial activity of primaquine and of exogenous xanthine oxidase has been found to be increased by pretreatment with the SOD inhibitor diethyldithiacarbamate [7, 145, 166, 181].

Decreased $\cdot OH$ and $O_2^-\cdot$ production was found in the phagocytes of leprotic patients, while the SOD level was higher than in the controls. The surplus SOD must have derived from the SOD of the phagocytosed *Mycobacte-*

rium. The defective respiratory burst of phagocytes, which is due to intrinsic or elevated SOD content, must play a decisive role in the insufficiency of the defence mechanisms in leprosy. Clofazimine, an effective leprostatic, probably acts as an iron chelating agent increasing phagocytic ·OH production [403].

20.1 Summary

Free radical reactions play an important part in the virulence of parasites and bacteria and in the body's defence against these microorganisms. The virulence of an organism depends on how much it activates the respiratory burst of phagocytes and on the antioxygen defence of the species. Active respiratory burst and poor antioxidant protection result in low virulence and vice versa. A number of antiparasitic drugs rely for their effect on free radical reactions.

21 Hormones and Free Radical Reactions

Corticosteroids are antioxidants and inhibit free radical reaction. Their LPO-decreasing action has been demonstrated in vitro. Corticosteroids inhibit phospholipase A_2 activation, and thus arachidonic acid release and the early steps of arachidonic acid metabolism. This mechanism may have a part in the anti-inflammatory action of corticosteroids. Increased susceptibility to infections, which is a side effect of corticosteroids, may be due to their inhibitory effect on the H_2O_2 production of mononuclear phagocytes. Nevertheless, not all corticosteroids have this side effect, which also depends on their ability to cross the cell membrane [253, 441, 520, 115].

Oestrogens, too, are potent antioxidants [115]. Thyroid hormones have been suggested to have an important role in the control of oxidant-antioxidant balance, although the pertinent data are not unequivocal. The antioxidant action of thyroxine and reverse T_3 has been demonstrated, whereas other authors found that the administration or endogenous overproduction of thyroid hormones produced a prooxidant state indicated by increased CuZnSOD activity. Thyroid hormones also increased oncogenic transformation, which could be inhibited with SOD. The antioxidant action of thyroid hormones does not exclude their producing a prooxidant state, since this effect may be due to an increase of the basal metabolism - which is a well-known effect of these hormones - leading to increased univalent leakage [433, 558].

21.1 Summary

Hardly anything is known about the connection between hormones and free radical reactions. Reliable data are only available as regards the antioxidant effect of corticosteroids.

22 Ophthalmological Aspects

The aetiological role of free radical reactions in the development of cataracts has been proved. The MDA content of rat lenses incubated with light was six times that of controls kept in the dark and with freshly removed lenses. SOD, CAT and ascorbate inhibited the photoperoxidation of the lens. Increased MDA was also found in human senile cataracts and in experimental cataracts. MDA forms cross-links with the aminophospholipids of the lens: a phosphatidylethanolamine-MDA-phosphatidylserine adduct was demonstrated by thin-layer chromatography. Other antioxidants, e.g. vitamin E and centrophenoxine were also found to have a protective effect in vivo in the experimental cataract of rabbits and also in vitro. Crouch measured concurrently the SOD level by an immunological technique and SOD activity by biochemical means, and found an age-dependent decrease in enzyme activity, while the amount of the immunologically reactive enzyme did not change. This indicates that the antioxidant protection of the lens decreases with age. Cataract frequency increases in the aged, in diabetes (in which free radical reactions also play a role), in herbicide poisoning caused by agents with a diquat/paraquat structure, and in repair deficiency syndromes. Senile cataract is caused by a slow photoperoxidation which is due to chronic light exposure. This probably leads to ROI release and LPO in the lens. The function of ascorbate which is present in the aqueous humour has not been established. The free radical theory of cataract formation explains the physiological protective function of the antioxidant ascorbate. In India for example, where there is exposure to excessive sunlight and the diet contains little vitamin C, the incidence of cataract is very high [35, 36, 96, 518, 561, 562].

Corneal endothelial lesion frequently occurs in inflammatory diseases of the eye. This process may be mediated, at least partly, by ROIs produced as a result of the function of cells participating in the inflammatory reaction (phagocytes). CAT was found to inhibit experimental damage of the corneal endothelium induced with ROI-generating systems. Its use, therefore, may be justified in the palliative treatment of inflammations of the eye [268, 269].

22.1 Summary

The free radical theory of the photoperoxidative mechanism of senile cataract and cataracts associated with drugs and diseases is well supported by experimental evidence and furnishes the most up-to-date explanation of cataractogenesis.

23 Free Radical Reactions in Muscular Dystrophy

The participation of free radicals in the development of primary muscular diseases has been indicated by several data, but their role is not clear. Besides other symptoms, deficiency of selenium and vitamin E causes muscular dystrophy in animals (cat, sheep, mice) [322, 412]. Increased levels of TBA reactive products, lysosomal enzymes and antioxidant enzymes have been described in dystrophic chickens. All these changes indicate the presence of oxidant stress. The level of serum and α-cysteine proteinase inhibitor (α-CPI) increased in vitamin E-deficient rats suffering from muscular dystrophy. The activity of the tissue lysosomal enzyme proteinase increases in muscle degeneration, inducing a compensatory increase in the activity of α-CPI, which inhibits proteinase. In addition to vitamin E, ethoxyquin administration was found to prevent the morphological changes and the growth of α-CPI activity - almost as actively as vitamin E - without, however, saving any vitamin E [227, 369, 370]. The chickens with muscular dystrophy are a good model of this condition in humas because of the similar histological changes. The data available refer to chickens in which muscular dystrophy had run its full course and it could not, therefore, be decided whether the prooxidant state is a primary cause of the disease, or whether it develops secondarily as a result of tissue damage or due to ROIs released during phagocytosis in the muscle.

Therefore, Mizuno studied these parameters 1-2 weeks after hatching, when the morphological changes are hardly apparent, and no phagocytes have yet been attracted to the muscle. Increased activity of antioxidant enzymes (CuZnSOD, CAT, GP, glutathione reductase) and an increased level of TBA reactive products were found in this case, too. This indicates the participation of free radical reactions in the aetiology of the disease, besides basic genetic factors. The activity of antioxidant enzymes and the level of TBA reactive products gradually decreased in the controls, while in the dystrophic chickens they persisted on a high level. The mechanism whereby the prooxidant state causes muscular dystrophy is not known. The decrease or loss of activity of some enzymes may lead to inhibition of the SH groups of enzymes and proteins, and the damaging effect of the subsequent LPO may also be involved [369, 370].

23.1 Myotonic Dystrophy

Decreased serum selenium levels have been found in myotonic dystrophy (MD). This is a chronic disease of unknown aetiology, and although its progression is slow, it cannot be influenced by therapy. The first reports on successful selenium therapy of MD have been published by Swedish authors. Long-term administration (13 months) of selenium resulted in considerable objective and subjective improvement, the muscular strength increased and the pathological EMG changes regressed [411, 412].

23.2 Duchenne-type Muscular Dystrophy

Primary changes of the muscle membrane probably play an important role in the aetiology of muscular dystrophies. In the Duchenne-type dystrophy changes have been demonstrated in the membrane and also in several membrane-bound enzymes. Lysosomol activation is another standard change, which is probably elicited by LPO. The level of the protective enzymes and of LPO was investigated in human dystrophies. LPO was found to be increased. An increase was also found in GP and CAT, whereas SOD was unchanged. This indicates the tendency of the tissues to suppress LPO, which, however, failed for some reason [291, 361, 524].

23.3 Summary

Selenium and vitamin E deficiency causes muscular dystrophy in animals. Biochemical changes indicating the presence of oxidant stress were found in dystrophic animals. The prooxidant state probably plays an aetiological role in dystrophy and is not secondary to tissue damage. In human myotonic dystrophy and Duchenne-type dystrophy free radical reactions may play a role, but this has not been proved. Antioxidant selenium therapy has been reported to be beneficial in myotonic dystrophy.

24 Free Radical Reactions in Pyelonephritis, Uraemia and Urological Diseases

Pyelonephritis is usually caused by ascending bacterial infection of the kidney and leads to its scarring and contraction. Scarring is unlikely to be caused by autoimmune mechanisms. It also occurs in the absence of cytotoxic cells (T-cells). If the acute inflammation is cured with antibiotics, the development of chronic pyelonephritis can be prevented. When animals were made neutropenic, the renal tissue was not damaged by bacterial infection. These data suggest that neutrophils have a role in the development of the disease. Persisting bacteria activate the respiratory burst, chemotactic factor is produced, and the inflammatory mechanism damaging the tubule cells is triggered. The process becomes chronic owing to the persisting bacteria and the chronic activation of neutrophils leads to fibrosis. In the monkey model of ascending pyelonephritis, SOD mitigated the inflammatory response and the damage of renal tissue, but did not prevent it, indicating that other factors besides free radicals are involved. SOD treatment might be a useful means of causal therapy [460].

Accumulation of TBA reactive substances was found in the blood of uraemic patients; the majority corresponded to MDA. Anaemia in chronic renal insufficiency is attributed to erythropoietin deficiency, inhibition of erythropoiesis and heme synthesis, and also to the shorter life span of erythrocytes, which is due to circulating plasma factors. MDA is one of the possible plasma factors. The incorporation of MDA into the red cell membrane and its consequences, i.e. changed cation permeability and decreased deformability, were described in Sect. 19. The erythrocytes of uraemic patients show both of these changes. In addition, decreased SOD activity and GSH level were found in the erythrocytes in uraemia. These were regarded as the secondary consequences of LPO. It has been suggested that MDA might be a uraemic toxin [300, 560].

SOD therapy has been tested in several urological diseases. Excellent results were achieved in irradiation cystitis and in other types of chronic cystitis, and in hydrocele. The effect was superior to that of earlier treatment and no considerable side effects were noted. In chronic prostatitis the results were no better than those achieved by traditional therapy. SOD therapy was also initiated in plastic induration of the penis and the early results were favourable. SOD therapy in these diseases is only palliative, it is aimed at the mitigation of inflammation [339].

24.1 Summary

The involvement of free radical reactions via the activation of the respiratory burst of phagocytes by persisting bacteria has been suggested in pyelonephritis. Accumulation of TBA reactive substances was shown in the blood of uraemic patients. LPO products may play the role of uraemic toxins and they probably also participate in the development of nepropathic anaemia. SOD proved to be a useful palliative agent in a number of urological diseases. The role of free radicals in the pathogenesis of these diseases has not been proved.

25 Other Diseases: Down's Syndrome

Down's syndrome is characterized by mental handicap, accelerated ageing, increased tumour frequency and immune dysfunction, among other manifestations. The CuZnSOD level is elevated owing to the gene dose effect, because this enzyme is coded for by chromosome 21. At the same time, the mitochondrial MnSOD level is lower. Increased GP concentration has been found in the patients' erythrocytes, which is probably due to a compensatory mechanism to neutralize H_2O_2 produced due to increased CuZnSOD activity. In the brain, however, the level of enzymes neutralizing H_2O_2 (GP and CAT) is low and probably cannot be induced either, or only slightly. This is shown by the normal cerebral GP activity of fetuses with Down's syndrome. Mental handicap, therefore, might be explained by the peroxidative damage of the brain. Increased LPO could indeed be demonstrated in Down's syndrome patients. The other changes mentioned can also be attributed to pathological free radical reactions. The connection between SOD level and mental handicap was discussed in Sect. 5.4 [52, 53, 309, 404].

25.1 Summary

Pathological free radical reactions which are due to a decreased mitochondrial MnSOD level probably play a role in the development of the changes characteristic of Down's syndrome.

26 The Role of Free Radical Reactions in Photobiology

Sunshine, like oxygen, may also have harmful effects irrespective of its important biological role. It may cause skin burn, DNA damage, skin cancer (basal cell carcinoma, epithelial melanoma), premature ageing of the skin and cataracts. The human skin is affected by wavelengths ranging from 290 to 720 nm (UV-A, UV-B, visible light). Free radical reactions participate in the development of all these lesions.

Sunburn of the skin is a complex inflammatory reaction in the course of which free radicals, lipid peroxides and, secondarily, PG are produced in the skin. Probably DNA is the primary acceptor of light energy.

Normal suntan develops in two phases. Immediate pigmentation, which is due to the oxidation of preformed melanin (a free radical reaction is also involved, see Cap. III Sect. 4.2), is followed by melanogenesis, i.e. the formation of new pigment within 3-4 days. Free radicals play a role in this phase, too, by increasing the activity of the tyrosinase enzyme.

Degenerative lesions of the skin (wrinkles, telangiectasia, keratosis) develop in those who are constantly exposed to sunshine. These changes correspond to the premature ageing of the skin. The frequency of skin cancer is also higher in such people. Free radicals are involved in this process, too.

Photosensitivity is the abnormal sensitivity of skin to light. Two types are distinguished:

1. Phototoxicity may develop in normal individuals as a result of concomitant exposure to the eliciting agent and to sunshine. It has no allergic component.
2. Photoallergy only develops in sensitive individuals. In this type allergic mechanisms play a mayor role.

Free radicals have an important part in phototoxicity. Phototoxicity reactions occurring in the presence of molecular oxygen are called photodynamic reactions. The sensitizing agent may act directly without oxygen radical formation (e.g. psoralen, which is converted to free radical by light absorption, produces ROIs in the presence of O_2 by releasing the energy it has acquired directly; porphyrin, see Sect. 18). The photoprotective and, at the same time, scavenger role of melanins is indicated by the increased sensitivity to this light-induced damage in people with white skin and hypopigmentation [345, 424].

26.1 Summary

The role of free radical reactions has been proved in the development of the noxious effects of sunshine (sunburn, DNA damage, premature ageing of the skin, cataract formation). Radicals may also participate in photocarcinogenesis. Melanin protects against sunshine owing to its scavenger action, but it may also be the source of ROI production. Free radical mechanisms also play a role in phototoxicity reactions.

Chapter VII

Defence Against Free Radical Reactions

1 Minimizing Exposure

Chiefly preventive measures belong under this heading, such as avoiding ionizing radiation and toxic chemicals, and the intake of an appropriate diet.

2 Antioxidants

The mechanism of effect of synthetic and natural antioxidants was described in Chap. II Sect. 3. Here, we deal with synthetic antioxidants and with the therapeutic application of both types of antioxidants. As regards their effect, antioxidants may be: (a) weak, acting in the period of initiation without decomposing hydroperoxides (e.g. vitamin E); (b) medium, having a moderate hydroperoxide decomposing action (e.g. biphenylamine derivatives); and (c) strong, acting in all phases of oxidation and decomposing hydroperoxides vigorously (e.g. BHT, EMQ, MTDQ, see later discussion) [25].

The best-known synthetic antioxidants include butylated hydroxyanisole (BHA), butylhydroxytoluene (BHT), ethoxyquin (EMQ), mercaptoethanolamine (MEA), dihydroquinoline-type antioxidants (MTDQ, MTDQ-DS), D-penicillamine, Catergen, lipoic acid, levamisole, CCA and Legalon.

The use of strong synthetic antioxidants in diseases caused by pathological free radical reactions is excluded by their toxic effects. Legalon, MTDQ and MTDQ-DS - the latter two developed by Hungarian researchers - are exceptions in this respect. Teratogenic and mutagenic side effects of some antioxidants have also been reported.

Retinoids (vitamin A precursors and derivatives). The anticarcinogenic action of vitamin A and its derivatives have received increased attention recently. They are used in the prevention and treatment of squamous cell epithelial carcinoma, because they inhibit metaplasia of the squamous epithelium. Local application of retinoic acid causes regression of irradiation keratosis and basal cell carcinoma. Oral doses produce regression of bladder papillomas and oral and laryngeal leukoplakia [610, 611]. Bjelke and co-workers found a correlation between decreased uptake of vitamin A and lung cancer. Therefore, the protective effect of vitamin A in smokers has been suggested. A negative correlation has been found between cancer frequency and vitamin A level. Other authors failed to find such a relation and suggested that the higher than average consumption of β-carotene may in this case act, not by being converted to vitamin A, but perhaps as a singlet oxygen scavenger. Retinoids had no significant effect in definitive cancers, but they may be used for preventing carcinogenesis and for the inhibition of further malignant transformation in cases where initiation has taken place. Vitamin A derivatives have been produced which are metabolized more rapidly, are less toxic and more suitable for therapeutic use. Isotretinoin (13-cis-retinoin acid) and etretinate (aromatic retinoid) are such vitamin A derivatives. Differentiation could be induced with retinoic acid in cultures of acute promyelocytic leukaemia cells and also in vivo in patients resistant to cytostatic therapy. Both the bone marrow and peripheral count normalized in these patients. Induction of differentiation could not be achieved in other types of myeloid leukaemia. Retinoids are used in dermatology, e.g. for the treatment of acne, psoriasis and disturbances of keratinization [3, 40, 45, 140, 170, 401, 414, 427, 453, 609, 611].

Vitamin E is a weak, natural antioxidant (for its effect and use see the appropriate sections).

Glutathione (GHS) (for its mechanism of effect see Fig. 3). Tumour regression in response to GSH was found in rats with aflatoxin B_1-induced hepatic tumour.

Vitamin C. Its anticarcinogenic effect has been demonstrated in animal experiments. It is also used for cataract prevention.

D-penicillamine (*β, β*-dimethylcysteine) has a metal chelating action. Its copper complex, or the low molecular weight copper complex released by gradual decomposition, has SOD-like, $O_2^-\cdot$ dismutating action. As a thiol compound it provides an alternative pathway for peroxidation, which it thus inhibits. It has been used with success, e.g. in RA, scleroderma, cystinuria, Wilson's disease, primary biliary cirrhosis (PBC), haemochromatosis, retrolental fibroplasia (RLF), and hyperbilirubinaemias of neonates. The frequent side effects (e.g. granulocytopenia, thrombocytopenia, anaemia, disturbed taste sensation, myasthenic reaction, renal damage), however, call for caution in its use [114, 311].

SOD is a widely used, effective anti-inflammatory agent the favourable effect of which in several diseases has been described in the present work. It is especially effective in osteoarthritis, RA, plastic induration of the penis, irradiation cystitis and bladder ulcers. The bovine product orgotein can be used in human therapy. Recently an SOD has been produced with genetic recombination in *Escherichia coli*, the sequence of which is practically identical with that of the SOD of human erythrocytes. In intravenous administration the short half-life of the product constitutes a problem, but it is also effective in subcutaneous or intramuscular doses. A product with longer half-life resulted from its coupling with macromolecules (PEG, Ficoll). Therapeutic SOD acts in the extracellular space, it cannot cross the plasma membrane. SOD enclosed in liposomes is also capable of intracellular action, but its use is still in an experimental stage. SOD, unlike other potent antiphlogistic agents, has no side effects. The only exception is the anaphylactic reaction caused rarely by bovine SOD. The use of the human product eliminates this side effect.

Low molecular weight compounds with SOD activity include D-penicillamine, nonsteroid anti-inflammatory agents and the histidine-Cu^{2+} complex. Compared with SOD, they have the advantage of easily entering the cell because of their low molecular weight [177, 295, 319, 466].

MTDQ, MTDQ-DS. MTDQ (6,6-methylene-bis-2,2,3-trimethyl-1,2-dihydroquinoline) is a lipid-soluble dihydroquinoline derivative, which is neither toxic nor teratogenic. Its radiosensitizing action is unique among antioxidants. Similar doses of other antioxidants are radioprotective. It is a strong antioxidant, and is most effective at body temperature. It selectively accumulates in tumour tissues and oxygen is released in situ as a result of its interaction with peroxides present in the neoplastic tissue, which plays a major role in its radiosensitizing action. In cutaneous basal cell carcinoma its use allowed a 50% reduction of the radiation dose. It may also be used for enhancing the radiosensitivity of radioresistant tumours. We found a favourable effect of the drug in CCl_4-induced hepatic damage.

MTDQ-DS (2,2-dimethyl-4-methane sulphonic acid sodium-1,2-dihydroquinoline) is a water-soluble derivative of MTDQ and its effect is practically the same, with the exception that it acts in aqueous media. We found it effective in atherosclerosis and in galactosamine- and doxorubicin-induced hepatic damage. Since MTDQ and MTDQ-DS are the only nontoxic potent antioxidants known, they may, in the future, be expected to play an important role in the treatment of diseases caused by free radical reactions [25].

Catergen is a naturally occurring flavonoid with an antioxidant and membrane stabilizing action. It is practically nontoxic and has a hepatoprotective effect. It secondarily decreases microsomal drug metabolism in the P-450 locus. It protects against toxic hepatic damage caused by CCl_4, galactosamine and phalloidin. It may be used in alcoholic liver disease, acute viral hepatitis, chronic hepatitis and especially in chronic persisting hepatitis (CPH). It is not effective in chronic acitve hepatitis (CAH). Its use is especially advantageous in CPH, in which corticosteroids are ineffective or contraindicated because of their side effects. Catergen has been found to have an immune stimulating action, which is in harmony with its antioxidant effect [604].

Leuamisole is a known immune modulator and anthelmintic agent. The mechanism of the former action is not known. It was found to be a potent antioxidant in vitro in the hepatic microsomes of rat and it also inhibited LPO induced by X-irradiation. Yoshikawa and co-workers found it effective in the prevention of CCl_4-induced hepatic damage. In the serum of elderly patients it reduced the high level of TBA reactive substances. The immune modulator effect of antioxidants was described in Chap. VI Sect. 3. It is probable that levamisole is also an immune modulator by virtue of its antioxidant property [591]. Levamisole is metabolized in the organism to a biologically active compound with SH groups. Its antioxidant action may be due to its GSH-like effect [425].

Lipoic acid. There are data indicating the antioxidant action of this drug. This probably plays a role in its hepatoprotective effect. The antioxidant action may be due to the SH groups of its reduced form. It has been said to inhibit platelet aggregation [317, 332, 425].

Lobenzarite disodium is an antioxidant with immune modulator action. A double-blind study has proved its effectiveness in RA (see also Chap. VI Sect. 3).

Legalon is a membrane stabilizing strong antioxidant containing silymarine as its active agent. In animal experiments it was found effective against CCl_4-, galactosamine- and ethanol-induced hepatotoxicity and in *Amanita phalloides* poisoning. It was also found to reduce the hepatotoxic effect of whole body irradiation. It noncompetitively inhibits lipoxygenase and PG synthetase, which may play a role in its anti-inflammatory effect observed in chronic hepatitis. In clinical studies it was found to have a favourable effect in hepatitis, cirrhosis and toxic liver damage (CCl_4, galactosamine, ethanol, tuberculostatics, *Amanita phalloides* poisoning, anaesthesia and insecticides) [29, 37, 68, 98, 451].

Although the majority of the reports dealing with the effect of antioxidants refer to favourable effects, untoward effects have lately also been reported (carcinogenicity, mutagenicity, teratogenicity). Therefore, antioxidants should be used with great caution as long as there are unsolved controversies in respect of their safe use. The systemic application of large doses of antioxidants is, therefore, not recommended at present.

3 Summary

Pathological free radical reactions are combated, on the one hand, by minimizing exposure to the causative agents, or by their elimination (e.g. ionizing radiation, noxious chemicals, dietary hazards), and, on the other, with antioxidant agents used as both preventive and curative drugs. Strong synthetic antioxidants cannot be used in human therapy because their effective doses are toxic. MTDQ, MTDQ-DS and Legalon are potent nontoxic antioxidants and are, therefore, expected to play a major role in the therapy of human free radical diseases. Other antioxidants that seem to be useful in the therapy of these diseases include the vitamins A, E and C, glutathione, D-penicillamine, SOD, low molecular weight substances with SOD-like activity, Catergen, levamisole, lipoic acid and lobenzarite disodium. Untoward side effects (carcinogenicity, mutagenicity, teratogenicity) have lately also been reported. The systemic use of large doses is not recommended until all aspects have been clarified.

Epilogue

We hope that, beyond summarizing the basic facts concerning free radical reactions, we have succeeded in conveying to the reader our firm conviction that free radicals play a very important role in biology and medicine. This hopefully does not stem from the inevitable partiality that all researchers feel for their respective fields. The findings of our investigations not only aroused our enthusiasm, but also raised doubts as we found data on the participation of free radical reactions in an increasing number of diseases. Agents appearing to be universal pathogens and drugs regarded as panaceas are bound to be looked upon with suspicion. Apparently this is almost the case with free radical reactions on the one hand, and with antioxidants on the other. Why are free radical reactions involved in such a large number of seemingly independent diseases? One possible explanation is that pathological free radical reactions damage membranes as well as the genetic material. Since membranes are found in every cell and every organ, their damage affects many cells, organs and organ systems. On the other hand, one wonders why the participation of this process with such an extensive damaging effect can be restricted to diseases involving only a limited number of organs. The answer may lie in the different antioxidant protection of individual organs, in their varying exposure to oxidant stress, in the metabolic characteristics of agents inducing free radical reactions, and probably in a number of other, unknown factors. In some of the diseases mentioned free radical reactions do not play an aetiological role, but are only responsible for some of the manifestations. In experiments it is hard to decide whether increased LPO is secondary to tissue damage caused by any noxa, or whether it is the primary cause of the specific disease.

The high number of hypotheses referred to indicate that this is a relatively new field with many unsettled problems. Many years' research is needed until physiological and pathological free radical reactions can be regarded as having acquired their final place in medical knowledge.

In conclusion, let us call attention to the singularly original studies of Nobel prize winner Albert Szent-Györgyi, whose words we use as a motto. On the basis of his experimental results, he states no less than that the "living state" is the free radical state of proteins and that carcinogenesis is basically due to the inhibition of the desaturation of proteins, which would otherwise result in the production of their free radical state.

Glossary

Allylic hydrogen. Hydrogen atom bound to the α-methylene carbon atom.

Antioxidants. Substances providing protection against the damaging effects of reactive oxygen intermediates. Those found in living organisms are called natural antioxidants. They may act in different phases of lipid peroxidation. Antioxidants react stoichiometrically with reactive oxygen intermediates and with the products of lipid peroxidation: therefore, they must be present in adequate concentrations at the site of free radical reactions and they must be able to reach these sites. The same substance may behave either as an antioxidant or as a prooxidant, depending on its concentration and on the chemical milieu.

Ceroid. Lipopigment, the product of in vivo lipid peroxidation. Its structure resembles that of lipofuscin. It differs from lipofuscin in that it is not produced during normal ageing, but only in pathological processes.

Chemiluminescence. Light emission below the temperature of incandescence is called luminescence. In the course of in vitro and in vivo physicochemical reactions atoms may absorb chemical energy which they transform to light energy. During lipid peroxidation, especially in its terminal phase, molecules capable of luminescence (e.g. singlet oxygen) are produced. Chemiluminescence is proportional to the degree of in vivo lipid peroxidation, it is, therefore, suitable for the direct demonstration of free radical reactions in vivo.

ESR (electron spin resonance) = EPR (electron paramagnetic resonance. Method for demonstrating the weak magnetic field of the unpaired electrons of free radicals, furnishing information on the structure of free radicals. It is not sufficiently sensitive for testing biological samples because it only demonstrates free radicals in the 10^{-5}–10^{-7} M concentration range. For the demonstration of free radicals present in lower concentration and of rapidly decomposing radicals, indirect ESR spectroscopy (spin trapping) is used, in which information about the structure of free radicals is obtained from changes in the ESR spectrum when free radicals are added to substances with known spectra.

Fenton reaction. See Haber-Weiss reaction.

Free radicals. Molecules or fragments of molecules containing unpaired electrons in their outer orbitals. The unpaired electrons tend to form a pair, therefore free radicals are highly reactive and have a short half-life.

Free radical reactions. Chain reactions started by initiators causing hydrogen removal, which is followed by several steps of propagation in which the free radicals produced interact with one another and with the surrounding molecules, and terminated when the free radicals are converted to nonradical products.

Haber–Weiss reaction. Reaction in which the potent oxidant hydroxyl radical is produced from superoxide radical anion and hydrogen peroxide

$$O_2^-\cdot + H_2O_2 \rightarrow O_2 + OH^- + \cdot OH$$

In living organisms the reaction is very slow. It is, however, accelerated by metal catalysts, when it is called the Fenton reaction.

Initiation See free radical reactions.

Lipid peroxidation. Process in which a lipid free radical R· is formed by the effect of a free radical initiator. The free radical then reacts with molecular oxygen, producing a peroxy free radical ROO·. A chain reaction is started, the effect of which is not restricted to lipids, but may, under pathological conditions, damage neighbouring biomolecules (proteins, carbohydrates, nucleic acids). Moreover, by mediation of the stable and transiently stable products of LPO, remote molecules may also be damaged. The process remains localized under physiological conditions, when the reaction is terminated after a series of steps.

Lipofuscin. Age pigment accumulating in humans and animals during the physiological process of ageing. It has been proved to be the product of lipid peroxidation. Its structure resembles that of ceroid and its presence may be regarded as the sign of in vivo lipid peroxidation. Lipofuscin has been shown to correspond to residual body, the end product of lysosomal digestion, which is formed during the lipid peroxidation of the polyunsaturated fatty acids of subcellular membranes.

Malondialdehyde. One of the stable end products of lipid peroxidation and of prostaglandin synthesis, which, in fact, is controlled enzymatic lipid peroxidation.

α-Methylene carbon atom. Carbon atom of fatty acids next to the double bond. Its C–H bond is weakened by the neighbouring double bond and can, therefore, be regarded as partially activated. Therefore, in fatty acids it is the α-methylene carbon atom from which hydrogen is most easily removed as the first step of lipid peroxidation (initiation).

Oxygen-dependent killing. Phagocytes can kill their targets via oxygen-dependent as well as oxygen-independent mechanisms. Oxygen-independent mechanisms are mediated by lysozyme, cationic proteins and lysosomal enzymes for example. Oxygen-dependent mechanisms are characterized by the formation of reactive oxygen intermediates.

Photosensitivity. Increased light sensitivity of the skin. It has two types: phototoxicity, which is due to free radical reactions, and photoallergy caused by allergic mechanisms. Phototoxicity reactions requiring the presence of molecular oxygen are called photodynamic reactions.

Prooxidants. Substances facilitating the formation of reactive oxygen intermediates and lipid peroxidation.

PUFA (polyunsaturated fatty acid). Lipid constituents which are the most prone to lipid peroxidation because their α-methylene carbon atoms next to the double bond are partially activated. Initiation of lipid peroxidation occurs readily in PUFAs.

Reactive oxygen intermediates. Collective term used for oxygen free radicals and reactive products of oxygen which, although not free radicals themselves, resemble free radicals in their reactivity. They are formed from oxygen by reduction or excitation.

Respiratory burst. A coordinated series of metabolic reactions occurring in stimulated phagocytes providing reactive oxygen intermediates for the killing of the target cells.

Scavengers. Antioxidants inhibiting free radical reactions by removing the radicals.

Spin trapping. See ESR spectroscopy.

Superoxide-dependent plasma chemotactic activity. Neutrophil chemotactic activity of a plasma precursor under the influence of superoxide radical anion. It enhances the inflammatory reaction. The anti-inflammatory effect of superoxide dismutase is due to the inhibition of this activity.

Thiobarbituric acid (TBA) reaction. Assay based on the acidcatalysed decomposition of lipid peroxide during which MDA is formed. MDA reacts with TBA, producing red chromogen. The reaction is not suitable for the demonstration of hydroperoxides from which no MDA is produced, whereas a number of biological substances give false-positive reactions. The accuracy of the method is limited by the reactive nature of MDA, which enters into reaction with the neighbouring molecules in situ. Since the reaction demonstrates substances other than MDA, these should be referred to as TBA reactive substances.

Univalent leakage. Process in which superoxide radical anion is produced in small, but constant amounts during oxidative phosphorylation in the mitochondria utilizing 1%–2% of the electrons transferred to cytochrome oxidases. (Univalent because the superoxide anion is the result of the one-electron reduction of oxygen.)

References

1 Abdalla, P., Monteiro, H.P., Bechara, E.J.H. (1985) Erythrocytic SOD levels in schizophrenic and manic-depressive patients. Fourth international Conference on superoxide and superoxide dismutase, Rome, 1-6 September 1985, abstract p 197

2 Acosta, D., Combs, A.B., Ramos, K. (1984) Attenuation by antioxidants of Na K ATP-ase inhibition by toxic concentrations of isoproterenol in cultured rat myocardial cells. J. Mol. Cell Cadiol. 16: 281-284

3 Aglietta, M., Piacibello, W., Sanavio, F., Visconti, A. Gavosto, F. (1984) Retinoic acid enhances the growth of only one subpopulation of granulomonocyte precursors. Acta Haematol 71: 97-99

4 Albano, E., Poli, G., Chiarpotto, E., Biasi, F., Dianzani, M.U. (1983) Paracetamol-stimulated lipid peroxidation in isolated rat and mouse hepatocytes. Chem.-Biol. Internat. 47: 249-263, 1983

5 Aldrich, T.K., Fischer, A.B., Cadenas, E., Chance, B. (1983) Evidence for lipid peroxidation by paraquat in the perfused rat lung. J. Lab. Clin. Med. 101. 66-73

6 Alhanaty, E., Snyder, M., Sheetz, M.P. (1984) Glucose-6-phosphate-dehydrogenase-deficient erythrocytes have an impaired shape recovery mechanism. Blood, 63: 1198-1202

7 Allison, A.C., Eugui, E.M. (1982) A radical interpretation of immunity to malaria parasites. Lancet ii: 1431-1433

8 Al-Timini, D.J., Dormandy, T.L. (1977) The inhibition of lipid autooxidation by human coeruloplasmin. Biochem J. 168: 283-288

9 Ambani, L.M., Van Woert, M.H., Murphy, S. (1975) Brain peroxidase and catalase in Parkinson disease. Arch. Neurol. 32: 114-118

10 Anderson, T.W., Reid, D.B.W. A double-blind trial of vitamin E in angina pectoris. In: Vitamin E Biochemistry, nutritional requirements, and clinical studies. J. Clin. Nutr. 27: 117-1178

11 Asayama, K., English, D., Slonim, A.E., Burr, I.M. (1984) Chemiluminescence as an index of drug-induced free radical production in pancreatic islets. Diabetes 33: 160-163

12 Asmus, K.D. (1984) Pulse radiolysis methodology. Methods Enzymol 105: 167-178

13 Babior, B.M. (1978a) Oxygen-dependent microbial killing by phagocytes [First of two parts]. N. Eng, J. Med. 298: 659-668

14 Babior, B.M. (1978b) Oxygen-dependent microbial killing by phagocytes [Second of Two Parts]. N. Eng, J. Med. 298: 721-725

15 Babior, B.M. (1982) The enzymatic basis for O_2 production by human neutrophils. Can. J. Physiol Pharmacol 60 1353-1358

16 Bachmann, R. (1954) Die Nebenniere. In: Bargmann W (ed): Blutgefäß- und Lymphgefäßapparat. Innersekretorische Drüsen. Springer, Berlin Göttingen Heidelberg, pp 1-952 (Handbuch der mikroskopischen Anatomie des Menschen, vol 6, part 5)

17 Bachur, N.R., Gordon, S.L., Gee, M.U. (1979) A general mechanism for microsomal activation of quinone anticancer agents to free radicals. Cancer Res. 38: 1745-1750

18 Badwey, J.A., Karnovsky, M.L. (1980) Active oxygen species and the functions of phagocytic leukocytes. Annu Rev. Biochem. 49: 695-726

19 Ball, S.S., Weindruch, R., Walford, R.L. (1985) Antioxidants and immune function. In: 1st International Congress of Biomedical Gerontology, New York, 10-11 July, 1985, abstract 4

20 Banga, I. (1978) Elastin and elastase in atherosclerosis research. Medicina, Budapest (Aesulap series)

21 Bannon, M.J., Goedert, M., Williams, B. (1984) The possible relation of glutatione, melanin and 1-methyl-4-phenyl-1,2,5,6-tetra-hydropyridine (MPTP) to Parkinson disease. Biochem Pharmacol 33: 2697-2698

22 Barbeau, A. (1984) Manganese and extrapyramidal disorders. Neurotoxicology 5: 13-36

23 Barboriak, J.J., Shetty, K.R., El-Ghatit, A.Z., Kalbfleisch, J.H. (1982) Plasma high-density lipoprotein cholesterol and vitamin E supplements. In: Lubin B, Machlin LJ (eds) Vitamin-E: biochemical, hematological, and clinical aspects. Ann. N.Y. Acad. Sci. 393: 174

24 Bartosiewicz, G., Mézes, M. (1983) The plasma level of vitamin E and activity of enzymes controlling peroxidation in rheumatoid arthritis and osteoarthritis. Magy Rheumatol 24: 65-70

25 Bär Vilmosné, Pollák, Zs. (1978) Development and use in oncoradiological practice of a radiosenzitizing antioxidant. Thesis, Budapest

26 Bata, M., Jávor, T., Moron, F., Patty, I., Tóth, Gy., Zsoldos, T., Mózsik, Gy. (1983) Interrelationships between the membrane-bound ATP-dependent energy systems of the gastric mucosa and the cytoprotective effect of betacarotene on the HCl-induced gastric mucosal damage in rats. International symposium on the recent advances in gastrointestinal cytoprotection, Pécs, 30 September-1 October 1983

27 Bates, E.J., Lowther, D.A., Handley, C.J. (1984) Oxygen free-radicals mediate an inhibition of proteoglycan synthesis in cultured articular cartilage. Ann. Rheumat. Dis 43: 462-469

28 Beckman, J.S., Campbell, G.A., Hannan C.Jr. Karfias, C. and Freeman, B. (1985) Involvement of superoxide and xanthine oxidase in cerebral ischemic injury. Fourth international conference on superoxide and superoxide dismutase, Rome, 1-6 September 1985, abstract p 55

29 Benda, L., Dittrich, H., Ferenci, P., Frank, H., Wawalka, F. (1980) Zur Wirksamkeit von Silymarin auf die Überlebensrate von Patienten mit Leberzirrhose. Wien Klin Wochenschr 119: 240

30 Benedetti, A., Malvaldi, G., Fulceri, R., Comporti, M. (1984) Loss of lipid peroxidation as a histochemical marker for preneoplastic hepatocellular foci of rats. Cancer Re 44: 5712-5717

31 Beregi, E. (1981) The medical problems of ageing. Medicina, Budapest (Aesculap series)

32 Beretta, L., Gerli, G.C., Ferraresi, R., Agostoni, A., Gualandri, V., Orsini, G.B. (1983) Antioxidant system in sickle red cells. Acta Haematol 70: 194-198

33 Best, L., Malaisse, W.S. (1983) Phospholipids and islet function. Diabetologia 25: 299-305

34 Betts, W.H., Cleland, L.G., Gee, D.J., Whitehouse, M.W. (1984) Effects of D-penicillinamine on a model of oxygen-derived free radical mediated tissue damage. Agents Actions 14: 283-290

35 Bhuyan, D.K., Podos, S.M., Bhuyan, K.C. (1985) Antioxidants in prevention of oxidative damage to the lens and of cataract in vivo. Fourth international conference on superoxide and superoxide dismutase, Rome, 1-6 September 1985, abstract p 57

36 Bhuyan, K.C., Master, R.W.P., Bhuyan, D.K. (1985) Oxy radical-induced polymerization of amino phospholipide of lens. Fourth international conference on superoxide and superoxide dismutase, Rome, 1-6 September 1985, abstract p 58

37 Bindoli, A., Cavallini, L., Siliprandi, N. (1977) Inhibitory action of silymarin on lipid peroxid formation in rat liver mitochondria and microsomes. Biochem. Pharmacol. 26: 2405

38 Bisgaard, H., Groth, S., Madsen, F. (1985) Bronchial hyperreactivity to leucotriene D_4 and histamine in exogenous asthma. Br. Med. J. 290: 1468-1471

39 Bishop, J.M. (1983) Cellular oncogenes and retroviruses Annu Rev. Biochem 52: 301-354

40 Bjelke, E. (1975) Dietary Vitamin A and human lung cancer Int. J. Cancer 15: 561-565

41 Blake, D.R., Hall, N.D., Bacon, P.A., Dieppe, P.A., Halliwell, B., Gutteridge, J.M.C. (1981) The importance of iron in rheumatoid disease. Lancet ii: 1142-1144

42 Boda, D., Németh, I., (1984) Possibility of protection against oxidant stress effect by the inhibition of xanthine oxidase with drugs. In: Oxygen free radicals and tissue damage, Scientific session, Pécs, 10-11 January 1984

42a Bode, C., Goebell, J., Beckmann, D., Dölle, W. (1968) Lysosomale Enzyme in Serum bei Erkrankungen der Leber und Gallenwege. Verh. dtsch. Ges. inn. Med. 74: 244-250

43 Borel, J.P.: Radicaux libres oxygénés, tissu conjonctif et inflammation. Pathol Biol (Paris) 31: 8-10

44 Borello, S., Minotti, G. Galeotti, T. (1985) Factors influencing O_2 and t-Bu-OOH dependent lipid peroxidation of tumour microsomes. Fourth international conference on superoxide and superoxide dismutase, Rome, 1-6 September 1985, abstract p 59

45 Braitman, T.R., Collins, S.J., Keene, B.R. (1981) Terminal differentation of human promyelocytic leukemic cells in primary culture in response to retinoic acid. Blood 57: 1000-1004

46 Boveris, A., Cadenas, E., Chance, B. (1981) Ultraweak chemiluminescence: a sensitive assay for oxidative radical reactions. Fed Proc 40: 195-198 (Physiological significance of lipid Peroxidation FASEB Symposium)

47 Börzsönyi, M. (1981) Production of N-nitroso compounds from precursors. Thesis, Budapest

48 Braganza, J.M., Wickens, D.G., Cawood, P., Dormandy, T.L. (1983) Lipid peroxidation (free-radical-oxidation) products in bile from patients with pancreatic disease. Lancet ii: 375-379 (1981)

49 Brawn, K., Fridowich, I. (1981) DNA strand scission by enzymatically generated oxygen radicals. Arch. Biochem. Biophys. 206: 414-419

50 Brizzee, K.R., Eddy, D.E., Harman, D., Ordy, J.M. (1984) Free radical theory of aging: effect of dietary lipids on lipofuscin accumulation in the hippocampus of rats. Age 7: 9-15

51 Bromberg, Y., Piek, E. (1984) Unsaturated fatty acids stimulate NADPH-dependent superoxide production by cell-free system derived from macrophages. Cell. Immunol. 88: 213-221

52 Brookshank, B.W.L., Balázs, R. (1984) Superoxide dismutase, glutathione per-

oxidase and lipoperoxidation in Down's syndrome fetal brain. Develop. Brain Res. 16: 37-44

53 Brookshank, B.W.L., Balázs, R. (1983) Superoxide dismutase and lipoperoxidation in Down's syndrome fetal brain. Lancet i: 881-882

54 Bulkley, G.B. (1985) Role of oxygen-derived free radicals as mediators of postischemic injury: an overview. Fourth international conference on superoxide and superoxide dismutase, Rome 1-6 September, 1985, abstract L 29

55 Burton, G.W., Cheeseman, K.H., Ingold, K.U., Slater, T.F. (1983) Lipid antioxidants and products of lipid perixodation as potential tumour protective agents. Biochem. Soc. Trans. 11: 261-262

56 Burton, K.P., McCord, J.M., Ghai, G. (1984) Myocardial alterations due to free radical generation Am. J. Physiol. 246: H 776-H 783

57 Bus, J.S., Gibson, J.E. (1982-1983) Mechanisms of superoxide radical mediated toxicity. J. Toxicol. Clin. Toxicol. 19 (6,7): 689-697

58 Cadenas, E., Boveris, A., Chance, B. (1984) Low-level chemiluminescence of biological systems. In: Pryor wA (ed) Free radicals in biology, vol VI. Academic, New York, pp 211-244

59 Cahill, G.F., Jr., McDevitt, H.D. (1981) Insulin-dependent diabetes mellitus: the initial lesion. N. Engl. J. Med. 304: 1454-1465

60 Campbell, E.J., Senior, R.M., McDonald, S.A., Cox, D.L. (1982) Proteolysis by neutrophils. Relative importance of cellsubstrate contact and oxidative inactivation of proteinase inhibitors in vitro. J. Clin. Invest. 70: 845-852

61 Capel, I.D., Dorrel, H.H., Smallwood, A.E. (1983) The influence of cold-restraint stress on some components of the antioxidant defence system in the tissues of rats of various ages. J. Toxicol. Environ. Health 11: 425-436

62 Capel, I.D., Thornley, A.C. (1983) Lipoperoxide levels, glutathione status and glutathione peroxidase activity in liver and tumours of mice bearing the Lewis lung carcinoma. Cancer Biochem. Biophys. 6: 167-172

63 Capel, I.D., Thornley, A.C. (1982) Superoxide dismutase activity, coeruloplasmin activity and lipoperoxide levels in tumour and host tissues of mice bearing the Lewis lung carcinoma. Eur. J. Cancer Clin. Oncol. 18: 507-513

64 Carp, H., Janoff, A. (1979) In vitro suppression of serum elastase - inhibitory capacity by reactive oxygen species generated by phagocytosing polymorphonuclear leukocytes. J. Clin. Invest. 63: 793-797

65 Carpenter, M.P. (1981) Antioxidant effects on the prostaglandin endoperoxide synthetase product profile. Fed. Proc. 40: 174-178 (Physiological significance of lipid peroxidation FASEB Symposium)

66 Carrell, R.W., Winterbourn, C.C., Rachmilewitz, E.A. (1975) Activated oxygen and haemolysis. Br. J. Haematol. 30: 259-264

67 Cathcart, R., Schwiers, E., Ames, B.N. (1983) Detection of picomole levels of hydroperoxides using a fluorescent dichlorofluorescein assay. Anal Biochem. 134: 111-116

68 Cavallini, L., Bindoli, A., Siliprandi, N. (1978) Comparative evaluation of antiperoxidative action of silymarin and other flavonoids. Pharmacol. Res. Commun. 10: 133

69 Cellerino, R., Guidi, G., Perona, G. (1976) Plasma iron and erythrocytic glutathione peroxidase activity. A possible mechanism for oxidative haemolysis in iron deficiency anaemia. Scand. J. Haematol. 17: 111-116

70 Cerutti, P.A. (1985) Prooxidant states and tumor production. Science, 227: 375-381

71 Chambers, D.E., Parks, D.A., Patterson, G., Yoshida, S., Burton, K., Parmley,

L.F., McCord, J.M., Downey, J.M. (1983) Role of oxygen derived radicals in myocardial ischaemia. Fed. Proc. 47: 1093

72 Chambers, D.J., Brainbridge, M.W., Hearse, D.J. (1985) Free radicals and ischemic injury: allopurinol pretreatment or free radical scavengers improve myocardial protection during global ischemia and reperfusion. Fourth international conference on superoxide and superoxide dismutase, Rome, 1-6 September, 1985, abstract p 171

73 Chan, A.C., Pritchard, E.T., Choy, P.C. (1983) Differential effects of dietary vitamin E and antioxidants on eicosanoid synthesis in young rabbits. J. Nutr. 113: 813-819

74 Chan, P.H., Fishman, R.A., Schmidley, J.W., Chen, S.F. (1984) Release of polyunsaturated fatty acids from phospholipids and alteration of brain membrane integrity by oxygen-derived free radicals. J. Neurosci. Res. 12: 595-605

75 Chan, P.H., Schmidley, J.W., Fischman, R.A., Longar, S.M. (1984) Brain injury, edema, and vascular permeability changes induced by oxygen-derived free radicals Neurology, 34: 315-320

76 Chen, X.S. (1982) Selenium and Keshan disease. In: Lubin B. Machlin, L.J. (eds) Vitamin E: biochemical, hematological, and clinical aspects. Ann. NY Acad. Sci. 393: 224-225

77 Chia, L.S., Thompson, J.E., Moscarello, M.A. (1983) Changes in lipid phase behaviour in human myelin during maturation and aging. Involvement of lipid peroxidation. FEBS Lett 157: 155-159

78 Chiu, D., Lubin, B. (1979) Abnormal vitamin E and glutathione peroxidase levels in sickle cell anemia. Evidence for increased suspectibility to lipid peroxidation in vivo. J. Lab. Clin. Med. 94: 542–548

79 Chiu, D., Vichinsky, E., Yee, M., Kleman, K., Lubin, B. (1982) Peroxidation, vitamin E and sickle-cell anemia. In: Lubin B., Machlin, L.J. (eds) Vitamin E: biochemical, hematological, and clinical aspects. Ann. NY Acad. Sci. 393, 323-333, 1982

80 Chow, C.K.: Dietary vitamin E and cellular susceptibility to cigarette smoking. In: Lubin, B., Machlin, L.J. (eds) Vitamin E: biochemical, hematological, and clinical aspects. Ann. NY Acad. Sci. 393: 426-454

81 Clark, R.A., Klebanoff, S.J. (1977) Studies on the mechanism of antibody dependent polymorphonuclear leukocyte-mediated cytotoxicity. J. Immunol. 119: 1413-1418

82 Cochran, T.A., Stefauko, J., Moore, C., Sack, R.P. (1983) Cytotoxic free radicals in stress ulceration Surg. Forum 33: 159-161

83 Cohen, A.S. (1980) Amyloidosis. In: Harrison's principles of internal medicine, 9th ed. Mc Graw Hill, Kokagusha, pp 338-341

84 Cohen, G. (1984) Oxy-radical toxicity in catecholamine neurons. Neurotoxicology 5: 77-82

85 Collipp, P.J., Chen, S.Y. (1981) Cardiomyopathy and selenium deficiency in a two-year old girl. N. Engl. J. Med. 304: 1304-1305

86 Cornides, Á., Kiss, Á., Sréter, L., Vereckei, A., Toncsev, H., Fehér, J. (1983) Inhibition of doxorubicin-induced hepatic toxicity by new dihydroquinoline type antioxidant. In: FALK symposium no. 38, Mechanism of hepatocyte injury and death, Basel, 3-5 October 1983

87 Cornwell, D.G., Morisaki, N. (1984) Fatty acid paradoxes in the control of cell proliferation: prostaglandins, lipid peroxides, and cooxidation reactions. In: Pryor WA (ed) Free radicals in biology, vol VI. Academic, New York, pp 95-148

88 Corwin, L.M., Cordon, R.K. (1982) Vitamin E and immune regulation. In: Lu-

bin, B., Machlin, L.J. (eds) Vitamin E: biochemical, hematological, and clinical aspects. Ann. NY Acad. Sci. 393: 437-450

89 Cusins, M.J. (1980) Halothane hepatitis: what's new? Drugs 19: 1-6

90 Craddock, P.R., Fehr, J., Dalmasso, A.P., Brigham, K.L., Jacob, H.S. (1977) Pulmonary vascular leukostasis resulting from complement activation by dialyzer cellophane membranes. J. Clin. Invest. 59: 879-888

91 Crane, D. Masters, C. (1984) On the role of catalase in the oxidation of tissue fatty acids. Arch. Biochem. Biophys. 229 (1): 104-111

92 Crary, E.J., McCarty, M.F. (1984) Potential clinical applications for high-dose nutritional antioxidants. Med. Hypotheses 13: 77-98

93 Crippa, P.R., Mazzini, A. (1983) Involvement of superoxide ions in the oxidation of NADH by melanins. Physiol. Chem. Phys. Med. NMR 15: 51-56

94 Crivello, J.F., Hornsby, P.J., Gill, G.N. (1983) Suppression of cultured bovine adrenocortical zone glomerulosa cell aldosterone synthesis by steroids and its prevention by antioxidants. Endocrinology 113 235-242

95 Crouch, R.K., Gandy, S.E., Kimsey, G., Galbraith, R.A., Galbratih, G.M.P., Nuse, M.G. (1981) The inhibition of islet superoxide dismutase by diabetogenic drugs. Diabetes 30: 235-241

96 Crouch, R.K., Patrick, J., Goosey, J., Coles, W.H. (1984) The effect of age on corneal and lens superoxide dismutase. Current Eye Res. 3: 1119-1123

97 Cruz, F.S., Lopes, L.A.V., De Souza, W., Moreno, S.N.J., Mason, R.P., Docampo, R. (1984) The photodynamic action of rose bengal on Trypanosoma cruzi. Acta Tropica, 41: 99-108

98 Csomós, G. Thaler, H. (1983) Clinical hepatology. Springer, Berlin, Heidelberg, New York 1983

99 Cuollar, A., Escamilla, E., Ramirez, J. Chavez, E. (1984) Adriamycin as an inhibitor of 11-beta-hydroxylase activity in adrenal cortex mitochondria. Arch. Biochem. Biophys. 235 538-543

100 Cukier, J.O., Maglalang, A.C., Odell, G.B. (1979) Increased osmotic fragility of erythrocytes in chronically jaundiced rats after phototherapy. Acta Fed. Scand. 68: 903-909

101 Cunningham, M.L., Lokesh, B.R. (1983) Superoxide anion generated by potassium superoxide is cytotoxic and mutagenic to Chinese hamster ovary cells. Mutation Res. 121: 299-304

102 Cutler, R.G. (1984) Antioxidants, aging and longevity. In Pryor wA (ed) Free radicals in biology, vol VI. Academic, New York pp 371-428

103 Dahlén, S.E., Hedqvist, P., Hammarström, S., Samuelsson, B. (1980) Leukotriénes are potent constrictors of human bronchi. Nature 288: 484-486

104 Dale, D.A., Bulkley, G.B., Granger, D.N., Hamilton, S.R., McCord, J.M. (1982) Ischemic injury in the cat small intestine: role of superoxide radicals. Gastroenterology 82: 9-15

105 Dalsing, M.C., Grosfeld, J.L., Schiffler, M.A., Vane, D.W., Hull, M., Baehner, R.I., Weber, T.R. (1983) Superoxide dismutase: a cellular protective enzyme in bowel ischaemia. J. Surg. Res. 34: 589-596

106 Das, S.K., Rajagopalan, C.N. (1980) Superoxide dismutase, glutathione, peroxidase catalase and lipid peroxidation of normal and sickled erythrocytes. Br. J. Haematol. 44: 87-92

107 Davis, B.W., Rennard, S.I., Bitterman, P.B., Crystal, R.G. (1983) Pulmonary oxygen toxicity. Early reversible changes in human alveolar structures induced by hyperoxia. N. Engl. Med. J. 309: 878-883

107a Dák, Gy., Nékám, K., Müzes, Gy., Romány, A., Vereckei, A., Láng, I., Fehér, J. (1986) Effects of *in vivo* therapy with 4-amino-5-imidasolcarboxamid-phosphate (Aica-P) on cellular immunoreactivity. Proceedings of conference on free radical scavengers in medicin, Szeged, 1986

108 Deby, C., Pincemail, J., Deby-Dupont, G., Braquet, P., Goutier, R. (1984) Secondary chemical reactions mediated by cyclooxygenase. In: Thaler-Dao, H. et al. (eds) Incesanoids and cancer. Raven, New York

109 De Groot, H., Noll, T. (1983) Halothane hepatotoxicity: relation between metabolic activation, hypoxia, covalent binding, lipid peroxidation and liver cell damage. Hepatology 3: 601-606

110 Del Maestro, R.F., Planker, M., Arfors, K.E. (1982) Evidence for the participation of superoxide anion radical in altering the adhesive interaction between granulocytes Int. J. Microcirc. Clin. Exp. 1: 105-120

111 Del Maestro, R.F., Thaw, H.H., Björk, J., Planker, M., Arfors, K.E. (1980) Free radicals as mediators of tissue injury. Acta Physiol. Scand. Suppl. 492: 43-57

112 Del Maestro, R.F. (1980) An approach to free radicals in medicine and biology. Acta Physiol. Scand. Suppl. 492: 153-168

113 Demopoulos, H.B. (1973) The basis of free radical pathology. Fed. Proc. 32: 1859-1861

114 Demopoulos, H.B. (1973) Control of free radicals in biologic systems. Fed. Proc. 32 1903-1908

115 Demopoulos, H.B., Flamm, E.S., Pietronigro, D., Seligman, M.L. (1980): The free radical pathology and the microcirculation in the major central nervous system disorders Acta Physiol. Scand. Suppl. 492: 91-119

116 Demopoulos, H.B., Flamm, E.S., Seligman, M.L., Pietronigro, D.D., Tomasula, J., De Crescito, V. (1982) Further studies on free radical pathology in the major central nervous system disorders: effect of very high doses of methylprednisolone on the functional outcome morphology and chemistry of experimental spinal cord impact injury. Active oxygen and medicine symposium. Can. J. Physiol. Pharmacol. 60: 1415-1424

117 Deneke, S.M., Fanburg, B.L. (1980) Normobaric, oxygen toxicity of the lung. N. Engl. J. Med. 303 (2): 76-86

118 Dickinson, A.C., DeJordy, J.O., Boutin, M.G., Teres, D. (1984) Absence of generation of oxygen-containing free radicals with 4'-deoxy-doxorubicin, a non-cardiotoxic anthracycline drug. Biochem. Biophys. Res. Commun. 125: 584-591

119 Dillard, C.J., Downey, J.E., Tappel, A.L. (1984) Effect of antioxidants on lipid peroxidation in iron-loaded rats. Lipids 19: 127-133

120 Docampo, R., Moreno, S.N.J. (1984) Free radical intermediates in the antiparasitic action of drugs and phagocytic cells. In: Pryor WA (ed) Free radicals in biology, vol VI. Academic, New York, pp 243-288

121 Docampo, R., Moreno, S.N.J. (1984) Free radical metabolites in the mode of action of chemotherapeutic agents and phagocytic cells on *Trypanosoma cruzi*. Rev. Infect. Dis. 6: 223-238

122 Doll, D.C. (1983) Oxidative haemolysis after administration of docorubicin. Br. Med. J. 287: 180-181

123 Donaldson, J. (1984) Involvement of manganese in physiological and biochemical processes an overview. Neurotoxicology 5: 1-4

124 Donaldson, J., La Bella, F.S. (1984) The effects of manganese on the colinergic receptor in vivo and in vitro may be mediated through modulation of free radicals. Neurotoxicology 5: 105-112

125 Dormandy, T.L. (1978) Free radical oxidation and antioxidants. Lancet: 647-650

126 Dormandy, T.L. (1983) An approach to free radicals. Lancet ii: 1010-1014

127 Doroshow, J.H., Locker, G.Y., Myers, C.E. (1980) Enzymatic defenses of the mouse heart against reactive oxygen metabolites. Alteration produced by doxorubicin. J. Clin. Invest. 65: 128-135

128 Dyerberg, J., Bang, H.O., Moncada, S., Vane, J.R. (1978) Eicosapentaenoic acid and prevention of thrombosis and atherosclerosis? Lancet ii: 117-119

129 Eddy, D.E., Harman, D. (1977) Free radical theory of aging: effect of age, sex and dietary precursors on rat-brain docosahexanoic acid. J. Am Geriat. Soc. 25 (5) 220-229

130 Ehrenkranz, R.A., Ablow, R.C., Warshanw, J.B. (1982) Effect of vitamin E on the development of oxygeninduced lung injury in neonates. In Lubin, B., Machlin, L.J. (eds) Vitamin E: biochemical, hematological, and clinical aspects. Ann. NY Acad. Sci. 393: 452-465

131 Ehrenkranz, R.A., Bonta, B.W., Ablow, R.C., Warshaw, J.B. (1978) Amelioration of bronchopulmonary dysplasia after vitamin E administration. N. Engl. J. Med. 2299: 564-569

132 Elek, G. (1971) Investigation of free radicals in some biological systems. Thesis, Budapest

133 Eling, T.E., Ally, A.I. (1984) Pulmonary biosynthesis and metabolism of prostaglandins and related substances. Environ Health Perspect 55: 159-168

134 Emerit, I., Cerutti, P. (1981) Clastogenic activity from Bloom's syndrome fibroblast cultures. Proc. Natl. Acad. Sci. USA 78: 1868-1972

135 Emerit, I., Levy, A., Cerutti, P. (1983) Suppression of tumor promoter phorbol myristate acetate-induced chromosome breakage by antioxidant inhibitors of arachidonic acid metabolism. Mutat. Res. 110: 327-335

136 Emerit, I., Michelson, A.M. (1980) Chromosome instability in human and murine autoimmune disease: anticlastogenic effect of superoxide dismutase. Acta Physiol. Scand. Suppl. 492: 59-65

137 Emerit, I., Braquet, M., Congy, F., Clavel, J.P. (1984) Superoxyde dismutase érythrocytaire chez l'alcoolique chronique avec lésions hépatiques. Presse 13: 1277-1278

138 Emerit, I., Cerutti, P. (1984) Icosanoide and chromosome damage. In: Thaler-Dao et al. (eds) Icosanoids and Cancer. Raven, New York, pp 127-138

139 Englberger, W., Bitter-Suermann, D. Hadding, U. (1984) Involvement of phospholipid metabolism in the events of the oxidative burst of granulocytes. Agents Actions 15: 31-33

140 Epstein, F.H. (1984) Vitamin A and retinoids in health and disease, N. Engl. J. Med. 310: 1023-1031

141 Erdey-Gruz, T. (1973) The foundations of physical chemistry 4th edn.

142 Erdey-Gruz, T. (1973) Foundations of the structure of matter 2nd edn. Müszaki Könyvkiadó, Budapest, p 283

143 Erskine, K.J., Iversen, S.A., Davies, R. (1985) An altered ratio of 18:2 (9,11) to 18:2 (9,12) linoleic acid in plasma phospholipids as a possible predictor of preeclampsia. Lancet i: 554-555

144 Esterbauer, H., Lang, J., Zadravec, S., Slater, T.F. (1984) Detection of malonaldehyde by high-performance liquid chromatography. Method Enzymol 105: 319-328

145 Fairfield, A.S., Meshnick, S.R. (1984) Host superoxide dismutase incorporation by intraerythrocytic plasmodia. In: Malaria and the red cell. Liss, New York, pp 13-23

146 Fajer, A., Carneiro, H.B., Oria, H. (1955) Effect of vitamin E on the incidence of gastric ulcers in rats on simplified diets. Hospital (Brazil) 48: 543

147 Fantone, J.C., Ward, P.A. (1982) Role of oxygen-derived free radicals and me-

tabolites in leukocyte dependent inflammatory reactions. Am J. Pathol. 107: 397-418

148 Fantone, J.C., Kunker, S.L., Ward, P.A. (1982) Chemotactic mediators in neutrophil-dependent lung injury. Annu. Rev. Physiol. 44: 283-293

149 Farber, E. (1981) Chemical carcinogenesis. N. Engl. J. Med. 305 (23): 1379-1389

150 Fehér, J., Bär-Pollák, Zs., Sréter, L., Fehér, E., Toncsev, H. (1982) Biochemical markers in carbon tetrachloride- and galactosamine-induced acute liver injuries: the effects of dihydroquinoline-type antioxidants. Br. J. Exp. Pathol 63: 394-400

151 Fehér, J., Cornides, Á., Vereckei, A., Láng, I., Nékám, K., Perl, A., Gergely, P. (1984) Immunomodulatory effect of new radical scavengers. Drugs Exp Clin. Res. X(8-9): 549-562, 1984

151a Fehér, J., Jakab, L., Tenczer, J., Szilvási, I. (1978) Klinische Untersuchungen zur Leberschutzwirkung von 4-Amino-5-Imidazolkarboxamid und Orotsäure. Dtsch Z. Verdau Stoffwechselkr 37: 177-181

152 Fehér, J., Kiss, Á., Blázovics, A., Szondy, É., Toncsev, H., Mathiász, D., Gerö, S. (1985) Hypolipaemic effect and inhibition of lipid peroxidation with glunicate in rats treatment atherogenic diet. Drugs. Exp. Clin. Res. 11: 413-419

152a Fehér, J., Láng, I., Nékám, K., Csomós, G., Müzes, Gy., Deák, Gy. (1986) Effect of silibinin on the activity and expression of SOD in lymphocytes from patients with chronic liver disease. Free Radical Res. Commun. (to be published)

152b Fehér, J., Láng, L., Nékám, K., Gonzales-Cabello, R., Deák, Gy. and Gergely, P. (1986) Inhibitory effect of silibinin on lectin-induced lymphocyte blast transformation in vitro. Int. J. Clin. Phamacol. Res. (to be published)

153 Fehér, J., Pollák, Zs., Sréter, L., Fehér, E. (1983) Protection of hepatocyte membrane with antioxidant substances (cyanidanol-3, dihydroquinoline, etc.). In: Conference on the pathobiochemistry and pharmacology of liver, Mátrafüred, 23-24 September 1983

154 Fehér, J., Pollák, Zs., Sréter, L., Toncsev, H., Cornides, Á., Gogl, Á., Vereckei, A. (1984) Experimental models for the study of hepatoprotection. Acta Physiol. Hung. 64: 401-407

155 Fehér, J., Sulyok, S., Pollák, Zs., Toncsev, H., Cornides, Á., Blázovics, A., Szondy, É., Gerö, S. (1984) The effect of a recently developed dihydroquinoline-type radical scavenger in cholesterol induced hyperlipidaemia. In: Lenzi S., Descovich G.C. (eds) Arteriosclerosis and cardiovascular diseases. Editrice Compositori, Bologna, pp 87-91

156 Fehér, J., Toncsev, H., Cornides, Á., Sréter, L., Kiss, Á., Vereckei, A. (1982) Effect of free radicals and its inhibition with a new antioxidant in experimental atherosclerosis. Treci Jugoslovenski Sympozyum ohyperlipoproteinemijam, Novi Sad, 3-5 June 1982

157 Fehér, J., Toncsev, H., Fehér, E., Kiss, Á., Vasadi, Á. (1981) Lysosomal enzymes in sera and granulocytes of patient's with chronic liver diseases. Int. J. Tissue React. 3: 31-37

158 Fehér, J., Toncsev, H., Sréter, L., Cornides, Á., Kiss, Á. (1984) Effect of cyanidol-3 on lysosomal enzym activity of serum and granulocytes in chronic liver disease and active hepatitis. Int. J. Tissue React. 6: 75-80

159 Fehér, J., Toncsev, H., Sréter, L., Cornides, Á., Vereckei, A. (1983) Protection of hepatocyte membrane with antioxidant substances (cyanidanol-3, dihydroquinoline, etc.). Conference on the patho-biochemistry and pharmacology of liver, Mátrafüred, 23-24 September 1983

160 Fehér, J. (1980) The pathogenesis and prognosis of chronic active hepatitis. Thesis, Budapest

161 Ferencik, M., Schramek, S., Kazár, J. and Stefanovic, J. (1984) Effect of Coxiella burnetii on the stimulation of hexose monophosphate shunt and on superoxide anion production in human polymorphonuclear leukocytes. Act. Virol. 28: 246-250

162 Fernandez-Pol, J.A., Hamilton, P.D., Klos, D.J. (1982) Correlation between the loss of the transformed phenotype and an increase in superoxide dismutase activity in a revertant subclone of sarcoma virus-infected mammalian cells. Cancer Res. 42: 609-617

163 Fischer, L.J., Hanburger, S.A. (1980) Inhibition of alloxan action in isolated pancreatic islets by superoxide catalase, and a metal chelator. Diabetes 29: 213-216

164 Fisher, S.M. (1984) Arachidonate cascade and skin tumor promotion. In: Thaler-Dao et al. (eds) Icosanoids and cancer. Raven, New York, pp 79-90

165 Fischer, V. and Mason, R.P. (1984) Stable free radical and benzoquinone imine metabolites of an acetaminophen analogue. J. Biol. Chem. 259: 10284-10288

166 Fitch, C.D., Dutta, P., Kanjananggulpan, P., Chevli, R. (1984) Ferriprotoporphyrin IX: a mediator of the antimalarial action of oxidants and 4-aminoquinoline drugs. In: Malaria and the red cell. Liss, New York, pp 119-130

167 Flohé, L., Ötting, F. (1984) Superoxide dismutase assays. Methods in Enzymology 105: 93-104

168 Flohé, L., Günzler, W.A. (1984) Assays of glutathions peroxidase. Methods Enzymol 105: 114-121

169 Floyd, R.A., Lewis, C.A., Wong, P.K. (1984) High-pressure liquid chromatography-electrochemical detection of oxygen free radicals. Methods Enzymol 105: 231-237

170 Flynn, P.J., Miller, W.J., Weisdorf, D.J., Arthur, D.C., Brunning, R., Branda, R.F. (1983) Retinoic acid treatment of acute promyelocytic leukaemia: in vitro and in vivo observations. Blood 62: 1211-1217

171 Folkers, K. (1973) Relationships between coenzyme Q and vitamin E. In: Vitamin E: biochemistry, nutritional requirements, and clinical Studies. International Symposium, Minneapolis, Minnesota, 27-28 September 1973

172 Foote, C.S., Shook, F.C., Abakerli, R.B. (1984) Characterization of singlet oxygen, Methods Enzymol 105: 36-47

173 Fox, K.A.A., Bergmann, S.R., Sobel, B.E. (1985) Pathophysiology of myocardial reperfusion. Annu. Rev. Med. 36: 125-144

174 Frank, L., Yam, S., Roberts, R.J. (1978) The role of endotoxin in protection of adult rats from oxygen induced lung toxicity. J. Clin. Invest. 61: 269-275

175 Frank, L., Massaro, D. (1986) Oxygen toxicity. Am. J. Med. 69: 117-126

176 Freeman, B.A., Crapo, J.D. (1982) Biology of disease: free radicals and tissue injury. Lab. Invest. 47: 412-426

177 Fridovich, I. (1983) Superoxide radical: an endogenous toxicant. Annu Rev. Pharmacol. Toxicol. 23: 239-257

178 Fridovich, I. (1975) Superoxide dismutases. Annu. Rev. Biochem. 44: 147-159

179 Fridovich, I. (1978) The biology of oxygen radicals. Science 201: 875-880

180 Friedman, J., Cerutti, P. (1983) The induction of ornithine decarboxylase by phorbol 12-myristate 13-acetate or by serum is inhibited by antioxidants. Carcinogenesis 4: 1425-1427

181 Friedman, M.J. (1979) Oxidant damage mediates variant red cell resistance to malaria. Nature 280: 245-247

182 Fujimoto, Y., Maruta, S., Yoshida, A., Fujita, T. (1984) Effect of transition metal ions on lipid peroxidation of rabbit renal cortical mitochondria. Res. Commun. Chem. Pathol. Pharmacol. 44: 495-498

183 Füst, Gy. (1982) A newly recognized immunopathological mechanism occurring in haemodialysis and in diseases associated with complement activation. Orvosképzés 57: 105-116

184 Gabig, T.G., Babior, B.M. (1981) The killing of pathogenes by phagocytes. Annu. Rev. Med. 32: 313-326

185 Gadek, J.E., Fells, G.A., Crystal, R.G. (1979) Cigarette smoking induces functional antiprotease deficiency in the lower respirator tract of humans. Science 206: 1315-1316

186 Gandy, S.E., Buse, M.G., Sorenson, S.R.J., Crouch, R.K. (1983) Attenuation of streptozotocin diabetes with superoxide dismutase-like copper (II) (3,5-diisopropisalicylate) in the rat. Diabetologia 24: 474-484

187 Gandy, S.E., Buse, M.G., Crouch, R.K. (1982) Protective role of superoxide dismutase against diabetogenic drugs J. Clin. Invest. 70: 650-658

188 Gandy, S.R., Galbraith, R.A., Crouch, R.K., Buse, M.G., Galbraith, G.M.P. (1981) Superoxide dismutase in human islets of Langerhans. N. Engl. J. Med. 304 (25): 1547-1548

189 Garcia-Bunuel, L. (1984) Lipid peroxidation in alcoholic myopathy and cardiomyopathy. Med. Hypotheses 13: 217-231

190 Gardner, D.E. (1984) Oxidant-induced enhanced sensitivity to infection in animal models and their extrapolations to man. Hemisphere, New York, pp 423-439

191 Gardner, T.J., Stewart, J.R., Casale, A.S., Downey, J.M., Chambers, D.E. (1983) Reduction of myocardial ischemia injury with oxigen-derived free radical scavengers. Surgery 94: 423-427

192 Galdston, M., Levytska, V., Schwartz, M.S., Magnusson, B. (1984) Coeruloplasmin increased serum concentration and impaired antioxidant activity in cigarette smokers and ability to prevent suppression of elastase inhibitory capacity of alpha$_1$-proteinase inhibitor. Am. Rev. Respir. Dis. 129: 258-263

193 Galeotti, T., Borrello, S., Palombini, G., Masotti, L., Ferrari, M.B., Cavatorta, P., Arcioni, A., Stremmenos, C. and Zannoni, D. (1984) Lipid peroxidation and fluidity of plasma membranes from rat liver and Morris hepatoma 3924A, FEBS Lett 169: 169-173

194 Gay, B., Sanchez-Teff, S., Caravano, R. (1984) Ultrastructural localization of NADPH-oxidase activity in murine peritoneal macrophages during phagocytosis of Brucella Virchows Arch. [Cell Pathol.] 45: 147-155

195 Goremia, E., Corsare, C., Bonomo, H., Giardinelli, H., Pappalarde, P., Vanella, A., Sichel G. (1984) Emuelanins as free radicals trap and superoxide dismutase activities in amphibia. Comp. Biochem. Physiol. 79B: 67-69

196 Gerli, G.C., Beretta, L., Bianchi, M., Pelleggatta, A., Agostoni, A. (1980) Erythrocyte superoxide dismutase, catalase and glutathione peroxidase activities in beta-thalassaemia (major and minor). Scand. J. Haematol. 25: 87-92

197 Gerli, G.C., Beretta, L., Bianchi, M., Agostoni, A., Gualandr, V., Orsini, G.B. (1982) Erythrocyte superoxide dismutase, calatase and glutathine peroxidase in glucose-6-phosphate dehydrogenase deficiency. Scand. J. Haematol. 29: 135-140

198 Gerő, S., Blázovics, A., Szondy, É., Fehér, J. (1985) 44th European Atherosclerosis Group Meeting, Montreux, 14-16 March

199 Giardini, O., Cantani, A., Donfrancesco, A. (1981) Vitamin E therapy in homozygous beta-thalassaemia. N. Engl. J. Med. 305: 644

200 Glaino, J., Hartmann, S., Clemmesen, J., Jessen, K.E., Dam, M (1952) Studies on the role of lipoperoxides in human pathology. II. The presence of peroxidised lipids in the atherosclerotic aorta. Acta Pathol Microbiol Scand 30: 1-3

201 Goebel, K.M., Goebel, F.D., Schubotz, R. and Schneider, J. (1979) Hemolytic implications of alcoholism in liver disease. J. Lab. Clin. Med. 94: 123-32

202 Goebel, K.M., Storck, U. (1983) Effect of intra-articular Orgotein versus a corticosteroid on rheumatoid arthritis of the knees. Am J. Med. 74: 124-128

203 Goetzl, E.J. (1984) Asthma: new mediators and old problems. N. Engl. J. Med. 311: 252-253

204 Goetzl, E.J. (1980) Mediators of immediate hypersensitivity derived from arachidonic acid. N. Eng. J. Med. 303: 822-825

205 Goldstein, B.O., Harber, L.C.: Erythropoietic protoporphyria: lipid peroxidation and red cell membrane damage associated with photohemolysis. J. Clin. Inest. 51: 892-902

206 Goldstein, I.M., Roos, D., Kaplan, H.B., Weissman, G. (1975) Complement and immunoglobulins stimulate superoxide production by human leukocytes independently of phygocytosis. J. Clin. Invest., 56: 1155-1164

207 Gordon-Smith, E.C., White, S.M. (1974) Oxidative haemolysis and Heinz body haemolytic anaemia. Br. J. Haematol. 26: 513-517

208 Goyal, N., Patro, I.K., Sharma, S.P. (1985) Effects of centrophenoxin (ANP-235) on the extractability of cardiac lipofuscin. In: 1st international congress of biomedical gerontology, New York, 10-11 July, abstract 98

209 Graham, D.G.: Catecholamine toxicity: a proposal for the molecular pathogenesis of manganese neurotoxicity and Parkinson's disease. Neurotoxicology 5: 83-96

210 Granger, D.N., Rutili, G., McCord, J.M (1981) Superoxide radicals in feline intestinal ischaemia. Gastroenterology 81: 22-29

211 Grankvist, K., Marklund, S., Täljedal, I. (1981) Superoxide dismutase is a prophylactic against alloxan diabetes. Nature 294: 158-160

212 Greening, A.P., Lowrie, D.B. (1983) Extracellular release of hydrogen peroxide by human alveolar macrophages: the relationship to cigarette smoking and lower respiratory tract infections Clin. Sci. 65: 661-664

213 Greenstock, C.L. (1984) Free-radical processes in radiation and chemical carcinogenesis. Adv. Radiation Biol. 11: 269-293

214 Griffin, M., Weiss, J.W., Leitch, A.G., McFadden, E.R.Jr., Corey, E.J., Austen, K.F., Drazen, J. (1983) Effects of leukotriene D on the airways in asthma. N. Engl. J. Med. 308: 436-439

215 Gross, S.J., Landaw, S.A., Oski, F.A. (1977) Vitamin E and neonatal hemolysis. Pediatrics 59: 995-997

216 Gryglewski, R.J. (1979) Prostacyclin and atherosclerosis - a hypothesis. In: Gottd AM Tr, Smith LC, Alle B (eds) Atherosclerosis V. Proceedings of the fifth international symposium. pp 762-765 Springer, Berlin Heidelberg New York

217 Gutteridge, J.M.C., Rowley, D.A., Halliwell, B., Westermarck, T. (1982) Increased nonprotein-bound iron and decreased protection against superoxide radical damage in cerebrospinal fluid from patients with neuronal ceroid lipofuscinosis. Lancet ii: 459-460

218 Haeger, K. (1974) Long-time treatment of intermittent claudication with vitamin E. In: Vitamin E: biochemistry, nutritional requirements, and clinical studies. Symposium. Am. J. Clin. Nutr. 27: 1179-1181

219 Haeger, K. (1982) Long term study of alfa-tocopherol in intermittent claudication. In Lubin B., Machlin L.J. (eds) Vitamin E: biochemical, hematological, and clinical aspects. Ann. NY Acad. Sci. 393: 269-374

220 Hägglöf, B., Marklund, S.L., Holmgren, G. (1983) CuZn superoxide dismutase, catalase and glutathione peroxidase in lymphocytes and erythrocytes in insulin dependent diabetic children. Acta Endocrinol. 102: 235-239
221 Halliwell, B., Gutteridge, J.M.C. (1984) Lipid peroxidation, oxygen radicals, cell damage and antioxidant therapy. Lancet i: 1396-1397
222 Ham, E.A., Egan, R.W., Soderman, D.D., Gale, P.H., Kvehl, F.A.Jr. (1979) Peroxidase-dependent deactivation of prostacyclin synthetase. J. Biol. Chem. 254: 2191-2194
223 Hammerschmidt, D.E., Greenberg, C.S., Yamada, O., Craddock, P.R., Jacob, H.S. (1981) Cholesterol and atheroma lipids activate complement and stimulate granulocytes. A possible mechanism for amplification of ischemic injury in atherosclerotic states. J. Lab. Clin. Med. 98: 68-77
224 Hammond, B., Kontos, H.A., Hess, M.L. (1985) Oxygen radicals in the adult respiratory distress syndrome, in myocardial ischemia and reperfusion injury, and in cerebral vascular damage. Can. J. Physiol. Pharmacol. 63: 173-187
225 Handin, R.I., Karabin, R., Boxer, G.J. (1977) Enhancement of platelet function by superoxide anion. J. Clin. Invest. 59: 959-965
226 Hansson, R., Jonsson, O., Lundtam, S., Petterson, S., Schersten, T., Waldenström, J. (1983) Effects of free radical scavengers on renal circulation after ischemia in the rabbit. Clin. Sci. 65: 605-610
227 Harada, N., Minakata, K., Takahashi, Y., Ishikawa, M., Masui, T. (1984) Effect of non-biological antioxidant ethoxiquin on serum alpha-cysteine proteinase inhibitor level of Vitamin E deficient rats. Nutr. Rep. Internat. 30: 1307-1318
228 Harman, D. (1984) Free radical theory of aging: the "free radical" diseases. Age 7: 111-131
229 Harman, D. (1969) Prolongation of life: role of free radical reactions in aging. J. Am. Geriatr Soc. 17: 721-732
230 Harman, D., Hendricks, S., Eddy, D.E., Seibold, J. (1976) Free radical theory of aging: effect of dietary fat on central nervous system function. J Am. Geriatr Soc. 24 301-307
231 Harman, D. (1972) Free radical theory of aging: dietary implications. Am. J. Clin. Nutr. 25: 839-843
232 Harman, D. (1982) Nutritional implications of the free radical theory of aging. J. Am. Col. Nutr. 1: 27-34
233 Harman, D. (1978) Free radical theory of aging: nutritional implications. Age 1: 145-152
234 Harman, D. (1981) The aging process. Proc. Natl. Acad. Sci. USA 78: 7124-7128
235 Harman, D. (1980) Free radical theory of aging: beneficial effect of antioxidants on the life span of male NZB mice; role of free radical reactions in the deterioration of the immune system with age and in the pathogenesis of systemic lupus erythematosus. Age 3: 64-73
236 Harman, D. (1977) Free radical theory of aging: effect of free radical reaction inhibitors on the immune response J. Am. Geriatr. Soc. 25: 400-407
237 Harman, D. (1980) Secondary amyloidosis and antioxidants Lancet ii: 593
238 Harman, D., Eddy, D., Noffsinger, J. (1976) Free radical theory of aging: inhibition of amyloidosis in mice by antioxidants possible mechanism. J. Am. Geriatr Soc. 24: 203-210
239 Hashimoto, Y., Ziff, M., Hurd, E.R. (1982) Increased endothelial cell adherence, aggregation and superoxide generation neutrophils incubated in systemic lupus erythematosus and Felty's syndrome sera. Arthritis Rheum 25: 1409-1418

240 Hassan, H.M., Bhatti, A.R., White, L.A. (1984) Superoxide dismutase, catalase and peroxidase in four strains of Neisseria meningitidis of different virulence. FEMS Lett. 25: 71-74

241 Hassan, H.M., Moody, C.S. (1982) Superoxide dismutase protects against paraquat-mediated dioxygen toxicity and mutagenicity: studies in Salmonella typhimurium. In: Active oxygen and medicine. Symposium. Can. J. Physiol. Pharmacol. 60: 1367-1373

242 Hassan, H.M., Fridovich, I. (1979) Intracellular production of superoxide radical and of hydrogens peroxide by redox active compounds. Arch. Biochem. Biophys. 196: 385-395

243 Hata, S., Tanemura, M., Ohsugi, Y., Nakamo, T., Kaiho, S., Mizuno, K., Matsunaga, I., Takagaku, Y., Shindo, M. (1983) An anti-oxidative and immunomodulating actions of new antiarthritic agent, CCA (Lobenzarit Disodium). J Pharmacobio dyn 6: s-56

244 Hearse, D.J., Manning, A.S., Downey, J.M., Yellon, D.M. (1985) Xanthine oxidaseta critical mediator of myocardial injury during ischemia and reperfusion. In: Fourth international conference on superoxide and superoxide dismutase, Rome, 1-6 September, 1985 abstract L 11

245 Hebbel, R.P. (1984) Erythrocyte autooxidation in sickle cell anemia. TIPS: 371-372

246 Heidrick, M.L., Hendricks, L.C., Cook, E.E. (1984) Effect of dietary 2-mercaptoethanol on the life span, immune system, tumor incidence and lipid peroxidation damage in spleen lymphocytes of aging $BC3F_1$ mice. Mech Ageing Dev 27: 341-358

247 Heikkila, R.E., Cohen, G. (1973) 6-hydroxydopamine: evidence for superoxide radical as an oxidative intermediate. Science 181: 456-457

248 Heinecke, J.W., Baker, L., Rosen, H., Chait, A. (1984) Superoxide generated by smooth muscle cells mediates metal ion catalyzed LDL modification. Circulation 70: 11-344

249 Herbel, R.P., Eaton, J.W., Balasingam, M., Steinberg, M.H. (1982) Spontaneous oxygen radical generation by sickle erythrocytes. J. Clin. Invest. 70: 1253-1259

250 Hermann, W.S.Jr. (1982) The effect of vitamin E on lipoprotein cholesterol distribution. In: Lubin B., Machlin L.J. (eds) Vitamin E: biochemical, hematological and clinical aspects. Ann. NY Acad. Sci. 393: 467-471

251 Hervei, S., Bors, Zs., Ugrai M-né, Malik, T.: Czappán, P., Bohár, A., Véli, M. (1983) Prevention of retrolental fibroplasia with the administration of water-soluble vitamin E. Orv. Hetil 124: 811-812

252 Hess, M.L., Hanson, N.H., Okabe, E. (1982) Involvement of free radicals in the pathophysiology of ischemic heart disease. In: Active oxygen and medicine, Symposium. J. Physiol. Pharmacol. 60: 1382-1389

253 Hess, M.L. and Manson, N.H. (1983) The paradox of steroid therapy. Inhibition of oxygen free radicals. Circ Chock 10: 205-213

254 Hess, M.L., Okabe, E., Poland, J., Warner, M., Stewart, J.R., Greenfield, L.J. (1983) Glucose, insulin, potassium protection during the course of hypothermic global ischaemia and reperfusion: a new proposed mechanism by the scavenging of free radicals. J. Cardiovas. Pharmacol 5: 35-43

255 Hess, M.L., Manson, N. (1984) Review article. Molecular oxygen: friend and foe. J. Mol. Cell Cardiol. 16: 969-985

256 Heys, A.D., Dormandy, T.L. (1981) Lipid peroxidation in iron-overloaded spleens. Clin. Sci. 60: 295-301

257 Hilf, R., Smail, D.B., Murant, R.S., Leakey, P.B. and Gibson, S.L. (1984) He-

matoporphyrin derivate-induced photosensitivity of mitochondrial succinate dehydrogenase and selected cytosolic enzymes of R3230AC mammary adenocarcinomas of rats. Cancer Res. 44: 1483-1488

258 Hittner, H.M., Godio, L.B., Rudolfph, A.J., Adams, J.M., Garcia-Prats, J.A., Friedman, Z., Kauz, J.A., Monaco, W.A. (1981) Retrolental fibroplasia: efficacy of vitamin E in a double-blind study of preterm infants. N. Engl. J. Med. 305: 1365-1371

259 Hitzig, W.H., Seger, R.A. (1983) Chronic granulomatosus disease, a heterogeneous syndrome. Hum. Genet. 64: 207-215

260 Hochstein, P., Jain, S.R. (1981) Association of lipid peroxidation and polymerization of membrane protein with erythrocyte aging. In: Physiological significance of lipid peroxidation. FASEB Symposium. Fed. Proc. 40: 183-188

261 Holland, M.K., Alvarez, J.G., Storey, B.T. (1982) Production of superoxide and activity of superoxide dismutase in rábbit epididymal spermatozoa. Biol Reprod 27: 1109-1118

262 Horneffer, P.J., Gardner, T.J. (1984) Oxygen free radical scavengers reduce infarct size on reperfusion. Circulation 70, Supp II: II-260

263 Hornsby, P.J., Crivello, J.F. (1983) The role of lipid peroxidation and biological antioxidants in the function of the adrenal cortex. Part. 1: A background review. Molec, Cell Endocrinol. 30: 1-20

264 Hornsby, P.J., Crivello, J.F. (1983) The role of lipid peroxidation and biological antioxidants in the function of the adrenal cortex. Part. 2: Molec Cell Endocrinol 30: 123-147

265 Hornstra, G. (1982) Prostanoids and their precursor fatty acids in the management of arterial thrombosis. In: Progress in pharmacology, vol 4/4. Fischer, Stuttgart, pp 75-97

266 Hrushesky, W.J.M., Olshefski, R., Wood, P., Meshnick, S., Eaton, J.W. (1985) Modifying intracellular redox balance: an approach to improving therapeutic index Lancet i: 565-567

267 Hubers, H. (1983) Iron overload: pathogenesis and treatment with chelating agents. Blut 47: 61-67

268 Hull, S.D., Green, K., Hampstead, D. (1984) Potassium superoxide induction of rabbit corneal endothelial cell damage. Curr Eye Res. 3: 1321-1328

269 Hull, D.S., Green, K., Thomas, L., Alderman, N. (1984) Hydrogen peroxide-mediated corneal endothelial damage. Invest. Ophthalmol. Vis. Sci. 25: 1246-1253

270 Igali, S. (1977) Environmental mutagenesis (in Hungarian). Actual problems of biology 11: 143-237

271 Imai, H., Werthessen, N.T., Subramanyah, V., Le Quesne, P.W., Soldway, A.H., Kanisawa, M. (1980) Angiotoxicity of oxygenated sterols and possible precursors. Science 207: 651-653

272 Imaizumi, S., Kayama, T., Suzuki, J. (1984) Chemiluminescence in hypoxic brain-the first report. Stroke 15: 1061-1065

273 Iwakami, M. (1965) Peroxides as a factor of atherosclerosis. Nagoya J. Med. Sci. 28: 50-66

274 Iwata, K., Shibuya, H., Ohkawa, Y., Inui, N. (1984) Chromosomal aberrations in V79 cells induced by superoxide radical generated by the hypoxanthine-xanthine oxidase system. Toxicol Lett. 22: 75-81

275 Jacobs, H.S. (1981) The role of activated complement and granulocytes in shock states and myocardial infarction. J. Lab. Clin. Med. 98: 645-653

276 Jain, S.K., Yip, R., Hoesch, R.M., Pramanik, A.K., Dallman, P.R., Shohet, S.B. (1983) Evidence of peroxidative damage to the erythrocyte membrane in iron deficiency. Am. J. Clin. Nutr. 37: 26-30

277 Jakubowski, A., Pick, E. (1983) The mechanism of action of lymphokines VII. Modulation of the action of macrophage migration inhibitory factor by antioxidants and drugs affecting thromboxane synthesis. Immunopharmacology 6: 215-229

278 Janoff, A., Carp, H., Lee, D.K., Drew, R.T. (1979) Cigarette smoke inhalation decreases alpha-l-antitripsin activity in rat lung. Science 2062: 1313-1314

279 Janoff, A. (1985) Elastase in tissue injury. Annu. Rev. Med. 36: 207-216

280 Janzen, E.G. (1986) A critical review of spin trapping in biological systems. In: Pryor WA (ed) Free radicals in biology, vol IV. Academic, New York, pp 116-154

281 Janzen, E.G. (1984) Spin trapping. Methods Enzymol 105: 188-198

282 Jennings, R.B., Sommer, H.H., Symth, G.A., Flack, H.A., Linn, H. (1960) Myocardial necrosis induced by temporary occlusion of coronary artery in the dog. Arch. Path. 70: 82-92

283 Joenje, H., Eriksson, A.W., Frants, R.R., Arwert, F., Houwen, B. (1978) Erythrocyte superoxide-dismutase deficiency in Fanconi's anaemia. Lancet i: 204

284 Johnson, L., Schaffer, D., Quinn, G., Goldstein, D., Mathis, M.J., Otis, C., Boggs, T.R.Jr. (1982) Vitamin E supplementation and the retinopathy of prematurity. In: Lubin B., Machil L.J. (eds) Vitamin E: biochemical, hematological, and clinical aspects. Ann. NY Acad. Sci. 393: 473-493

285 Johnson, R.A., Baker, S.S., Fallon, J.T., Maynard, E.P., Ruskin, J.N., Wen, Z., Ge, K., Cohen, H.J. (1981) An occidental case of cardiomyopathy and selenium deficiency. N. Engl. J. Med. 304: 1210-1211

286 Johnston, R.B., Lehmeyer, J.E. (1976) Elaboration of toxic oxygen by-products by neutrophils in a model of immune complex disease. J. Clin. Invest. 57: 836-841

287 Jolly, S.R., Kane, W.J., Bailie, M.B., Abrams, G.D., Lucchesi, B.R. (1984) Canine myocardial reperfusion injury. Circ. Res. 54: 277-285

288 Julicher, R.H.M., Sterrenberg, L., Koomen, J.M., Bast, A. and Noordhock, J. (1984) Evidence for lipid peroxidation during the calcium paradox in vitamin E-deficient rat heart. Naunyn-Schmiedebergs Arch. Pharmacol. 326: 87-89

289 Kahl, R. (1984) Synthetic antioxidants: biochemical actions and interference with radiation, toxic compounds, chemical mutagens and chemical carcinogenesis. Toxicology 33: 185-228

290 Kangas, J.A., Schmidt, K.M., Solomon, G.F. (1972) Effect of vitamin E on the development of stress-produced gastric ulceration in the rat. Am. J. Clin. Nutr. 25: 864-866

291 Kar, N.C., Pearson, C.M. (1979) Catalase, superoxide dismutase, glutathione reductase and thiobarbituric acid-reactive products in normal and dystrophic human muscle. Clin. Chem. Acta 94: 277-280

292 Karmazsin, L. (1980) Immunobiological events in the postnatal adaptation of permaturely born infants. Thesis, Debrecen

293 Karpen, C.W., Cataland, S., Odorisio, T.M., Panganamala, R.V. (1984) Interrelation of platelet vitamin E and thromboxane synthesis in typel diabetes mellitus. Diabetes 33: 239-243

294 Karpen, C.W., Pritchard, K.A.Jr., Arnold, J.H., Cornwell, D.G., Panganamala, R.J. (1982) Restoration of prostacyclin/thromboxane A_2 balance in the diabetic rat. Influence of dietary vitamin E. Diabetes 31: 947-951

295 Kennedy, A.R., Troll, W., Little, J.B. (1984) Role of free radicals in the initiation and promotion of radiation transformation in vitro. Carcinogenesis 5: 1213-1218

296 Kessel, D. (1984) Porphyrin localization: a new modality for detection and therapy of tumors. Biochem. Pharmacol. 33: 1389-1393

297 Khan, W.A., Grupta, A., Shanker, U., Pandya, K.P. (1984) Involvement of iron and free radicals in benzene toxicity. Biochem. Pharmacol 33: 2009-2012

298 Kibata, M., Higuchi, Y.: Serum alfa-tocopherol, coenzyme Q and thiobarbituric acid-reactive substance in acute myocardial damage and stroke. In: Lubin B., Machlin L.J. (eds) Vitamin E: biochemical, hematological, and clinical aspects. Ann. NY Acad. Sci. 393: 179-182

299 Kikugawa, K., Kosugi, H., Asakura, T. (1984) Effect of malonaldehyde, a product of lipid peroxidation on the function and stability of hemoglobin. Arch. Biochem. Biophys. 229: 7-14

300 Kishore, B.K., Geyo, F., Arakawa, M. (1983) Lipid peroxidation in uremia: malonaldehyde - a putative uremic toxin. IRCS Med. Sci. 11, 750-751

301 Kitahara, A., Yamazaki, T., Ishikawa, T., Camba, E.A., Sato, K. (1983) Changes in activities of glutathione peroxidase and glutathione reductase during chemical hepatocarcinogenesis in the rat. Gann 74: 649-655

302 Klebanoff, S.J. (1980) Oxygen metabolism and the toxic properties of phagocytes. Ann Intern Med 93: 480-489

303 Klebanoff, S.J., Foerder, C.A., Eddy, E.M., Shapiro, B.M. (1979) Metabolic similarities between fertilization and phagocytosis. Conservation of a peroxidatic mechanism. J. Exp. Med. 149: 938-953

304 Kleinerman, J. (1977) Some effects of nitrogen dioxide on the lung. Fed. Proc. 36: 1714-1718

305 Koller, E., Lang, J., Jürgens, G., Esterbauer, H. (1985) Fluorescence properties of native, 4-hydroxy-nonenal modified and autooxidized low density lipoprotein. In: Fourth international conference on superoxide and superoxide dismutase, Rome, 1-6 September 1985, abstract P 37

306 Kornbust, D.J., Barfknecht, T.R., Ingram, P. (1984) Effect of di(2-ethylhexyl)phalate on DNA repair and lipid peroxidation in rat hepatocytes and on metabolic cooperation in chinese hamster V-79 cells. J. Toxicol. Environ Health 13: 99-116

307 Krylov, V.I., Olekhnovich, V.M., Sorogin, P.S., Zhogin, S.V. (1984) Activies of phospholipases and antioxidants in bronchial secretion products of children with bronchopulmonary diseases. Vopr. Med. Khim. 30: 52-55

308 Kraut, E.H., Sagone, A.L. (1981) The effect of oxidant injury on the lymphocyte membrane and functions. J. Lab. Clin. Med. 98: 697-703

309 Kudrin, A.N., Kogan, A.J., Koroljov, U.V., Nyikolajev, Sz. M., Beszkrovnova, N.N. (1978) The role of lipid peroxidation in the pathogenesis of myocardial infarction and favorable effect of antioxidants sodium selenite and vitamin E (in Russian). Kardiologia 18: 115

310 Lakatos, L., Karmazsin, L., Oroszlán, Gy., Balla, Gy. (1983) Possibilities of defense against oxygen toxicity in the newborn. Orv. Hetil 125: 247-253

311 Lakatos, L. (1979) Treatment of hyperbilirubinaemia of the newborn with D-penicillinamine. Thesis, Debrecen

312 Lamour, Y., Davous, P. (1983) Démence de type Alzheimer: données récentes. Presse Médi 12: 1415-1420

313 Lands, W.E.M., Kulmaez, R.J., Marshall, P.J. (1984) Lipid peroxide actions in the regulation of prostaglandin biosynthesis. In: Pryor WA (ed) Free radicals in biology, vol VI. Academic, New York, pp 39-61

314 Lands, W.E.M. (1984) Radicals and peroxides modulate the enzymic synthesis of eicosanoids from polyunsatured fatty acids. In: Thaler-Dao H, et al. (eds) Icosanoids and cancer. Raven, New York pp 41-48

315 Lawrence, G.D., Cohen, G. (1982) Ethane exhalation as an index of in vivo lipid peroxidation, concentrating ethane from a breath collection chamber. Anal Biochem. 122; 283-290

316 Legha, S.S., Wang, Y.M., Mackay, B., Ewer, M., Hortobágyi, G.N., Benjamin, R.S., Ali, M.K. (1982) Clinical and pharmacologic investigation of the effects of alphatocopherol on adriamycin cardiotoxicity. In: Lubin, B., Machlin L.J (eds) Vitamin E: biochemical, hematological and clinical aspects. Anm. NY Acad. Sci. 393: 411-417

317 Lehninger, A.L. (1975) Biochemistry, 2nd edn. Worth, New York, p 347

318 Lehninger, A.L. (1975) Biochemistry, 2nd edn. Worth, New York, p 503

319 Lengfelder, E. (1984) Can anti-inflammatory drugs act as scavengers of oxygen radicals? Agents Actions 15: 56

320 Lenzhofer, R., Magometschnigg, D., Dudczar, R., Cerni, C., Bolebruch, C., Moser, K. (1983) Indication of reduced doxorubicin induced cardiac toxicity by additional treatment with antioxidative substances. Experientia 39: 62-64

321 Leslie, R.D.G. (1983) Causes of insulin dependent diabetes. Br. J. Med. 287: 5-6

322 Levander, O.A. (1982) Clinical consequences of low selenium intake and its relationship to vitamin E. In: Lubin, B., Machlin L.J (eds) Vitamin E: biochemical, hematological, and clinical aspects. Ann. NY Acad. Sci. 393: 224-225

323 Lew, D.L., Southwick, F.S., Stossel, T.P., Whitin, J.C., Simons, E., Cohen, H.J. (1981) A variant of chronic granulomatous disease: deficient oxidative metabolism due to a low affinity NADPH oxidase. New Engl.J.Med.305: 1329-1333

324 Lewis, G.P. (1983) Immunoregulatory activity of metabolites of arachidonic acid and their role in inflammation. Br Med. Bull. 39: 243-248

325 Lewis, K.O., Paton, A. (1982) Could superoxide cause cirrhosis? Lancet ii: 188-189

326 Lewis, P.J., Dollery, C.T. (1983) Clinical pharmacology and potential of prostacyclin. Br Med. Bull. 39: 281-284

327 Ley, K., Arfors, K.E. (1982) Changes in macromolecular permeability by intramuscular generation of oxygen-derived free radicals. Microvasc Res 24: 25-33

328 Lippman, R.D., Agren, A. (1982) A new method that investigates superoxide versus respiration endomitochondrially with a site-specific chemiluminescent probe. J. Biochem. Biophys. Methods 7: 71-80

329 Lippman, R.D. (1982) A new method that investigates superoxide versus respiration in vitro using bioluminescence and Sepharose-bound adenosine derivatives. J. Biochem. Biophys. Methods 6: 81-97

330 Little, J.B., Kennedy, A.R., Nagasawa, H. (1983) Involvement of free radical intermediates in oncogenic transformation and tumor promotion in vitro. In: Radioprotectors and anticarcinogenes. Academic, New York, pp 487-493

331 Loeper, J., Goy, J., Emerit, J., Rozensztajn, L., Jeny, C., Bedu, O. (1983) Etude des acides gras et de la péroxydation lipidique au cours de l'athérosclérose humaine. Sem. Hop. Paris, 59: 1657-1660

332 Losonczi, H., Pár, A., Pakodi, F., Nagy, I., Jávor, T. (1984) Changes in platelet function under the effect of antioxidants. In: Oxygen free radicals and tissue damage, scientific session, Pécs, 10-11 January 1984

333 Loven, D.P., Guernsey, D.L., Oberley, L.W. (1984) Transformation affects superoxide dismutase activity. Int. J. Cancer 33: 783-786

334 Lown, J.W., Sondhi, S.M., Yen, S.F., Plambeck, J.A., Peters, J.H., Acton, E.M. and Gordon, G.R. (1984) Biophysical studies permitting an approach to a sec-

ond generation of less cardiotoxic anthracyclines. Drugs Exp Clin. Res. X: 735-744
335 Lown, J.W. (1984) Ethidium binding assay for reactive oxygen species generated from reductively activated Adriamycin (Doxorubicin). Methods Enzymol. 105: 532-539
336 Lucchesi, B.R. (1984) Myocardial reperfusion injury: reduction by free radical scavengers. Fourth international conference on superoxide and superoxide dismutase, Rome, 1-6 September 1985, abstract L 13
337 Magovern, G.J.Jr., Bolling, S.F., Casale, A.S., Bulkley, B.H., Gardner, T.J. (1984) The mechanism of mannitol in reducing ischemic injury: hyperosmolarity or hydroxyl scavenger. Circulation 70, suppl. I: I91, I95
338 Manning, A.S., Bernier, M., Hearse, D.J. (1985) "Anti-free-radical" interventions and reperfusion-induced arrhytmias. In: Fourth international conference on superoxide and superoxide dismutase, Rome, 1-6 September 1985, abstract p 175
339 Marberger, H., Huber, W., Bartsch, G., Schulte, T., Swoboda, P. (1974) Orgotein: a new antiinflammatory metalloprotein drug evaluation of clinical efficacy and safety in inflammatory conditions of the urinary tract. Int. Urol. Nephrol. 6 (2): 61-74
340 Marcus, A.J. (1979) Pathways of oxygen utilization by stimulated platelets and leukocytes. Sem Hematol 16 (3): 188-195
341 Marklund, S.L. (1982) Human copper-containing superoxide dismutase of high molecular weight. Proc. Natl. Acad. Sci. USA 79: 7634-7638
342 Marklund, S.L., Westman, N.G., Lundgren, E., Roos, G. (1982) Copper and zinc-containing superoxide dismutase, manganese-containing superoxide dismutase, catalase and glutathione peroxidase in normal and neoplastic human cell lines and normal human tissues. Cancer Res. 42: 1955-1961
343 Martin, W.J. (1984) Neutrophils kill pulmonary endothelial cells by a hydrogen-peroxide-dependent pathway: an in vitro model of neutrophil-mediated lung injury Am. Rev. Respir Dis. 130: 209-213
344 Mason, R.P.: Assay of in situ radicals by electron spin resonance. Methods Enzymol 105: 416-422
345 Mason, R.P., Chignell, C.F. (1980) Free radicals in pharmacology and toxicology-selected topics Pharmacol Rev. 33 (4): 189-211
346 Mason, R.P. (1984) Spin trapping free radical metabolites of toxic chemicals. In: Spin labelling in pharmacology, Academic, New York pp 87-129
347 Mason, R.P., Chignell, C.F., Eling, T.E., Mottley, C.: Free radical formation during the peroxidase-catalyzed oxidation of/bi/sulfite/hydrated sulfur dioxide/
348 Mason, R.P., Josephy, P.D. (1985) Free radical mechanism of nitroreductase. In: Rickert D.E. (ed) Toxicity of nitroaromatic compounds, Hemisphere, New York, pp 121-140
349 Marklund, S.L. (1984) Extracellular superoxide dismutase and other superoxide dismutase isoenzymes in tissues from nine mammalian species. Biochem. J. 222: 649-655
350 Marklund, S.L. (1984) Extracellular superoxide dismutase in human tissues and human cell lines. J. Clin. Invest. 74: 1398-1403
351 Matkovics, B., Szöllösiné Dr., Varga, I (1976) Radicals generated from molecular oxygen. Current problems of biology. vol 7. Medicina, Budapest, pp 203-268
352 Matkovics, B. (1984) Chemistry of biological lipid peroxidation. In: Oxygen free radicals and tissue damage, scientific session, Pécs, 10-11 January 1984
353 Matsumoto, M., Wakasugi, H., Ibayashi, H. (1981) Serum vitamin E, lipid per-

oxide and glutathione peroxidase in patients with chronic pancreatitis. Clin. Chim. Acta 110: 121-125

354 Matthews, S.B., Campbell, A.K. (1984) Neutrophil activation after myocardial infarction. Lancet ii: 756-757

355 Mauderly, J.L. (1984) Respiratory function responses of animals and man to oxidant gases and to pulmonary emphysema. Hemisphere, New York, pp 345-361

356 McCay, P.B., King, M., Lai, E.K., Poyer, S.L. (1983) The effect of antioxidants on free radical production during in vivo metabolism of carbon tetrachloride. J. Am Coll. Toxicol. 2: 195

357 McCord, J.M., Wong, K., Stokes, S.H., Petrone, W.F., English, D. (1980) Superoxide and inflammation: a mechanism for the antiinflammatory activity of superoxide dismutase. In: Free radicals in medicine and biology symposium. Acta Phys. Scand. Suppl. 492: 25-30

358 McCord, J.M. (1985) Oxygen-derived free radicals in postischemic tissue injury. N. Engl. J. Med. 312: 159-163

359 McCord, J.M., Roy, R.S. (1982) The pathophysiology of superoxide: roles in inflammation and ischaemia. In: Active oxygen and medicine symposium. Can J. Physiol. Pharmacol. 60: 1346-1352

360 McCord, J.M., Fridovich, I. (1978) The biology and pathology of oxygen radicals. Ann Intern Med 89: 122-127

361 Mechler, F., Imre, S., Diószeghy, P. (1984) Lipid peroxidation in Duchenne-type muscular dystrophy. In: Oxygen free radicals and tissue damage, scientific session, Pécs, 10-11 January 1984

361a Mendez, J.C.B., Videla Senza, C.J. (1984) Biochemische Untersuchungen des Wirkmechanismus des Flavonoids silibinin-dihemisuccinat. Thesis, Santiago, Chile

362 Menzel, D.B. (1984) Ozone: an overview of its toxicity in man and animals. Hemisphere, New York, pp 1183-1203

363 Meredith, M.J., Reed, D.J. (1983) Depletion in vitro of mitochondrial glutathione in rat hepatocytes and enhancement of lipid peroxidation by adriamycin and 1,3-bis(2-chloroethyl)-nitrosurea(BCNU). Biochem. Pharmacol 32: 1383-1388

364 Mergens, W.S. (1982) Efficacy of vitamin E to prevent nitrosamine formation. In: Lubin, B., Machlin, L.J. (eds) Vitamin E: bioechemical, hematological, and clinical aspects. Ann NY Acad. Sci. 393: 61-69

365 Metz, S., Van Rollins, M., Strife, R., Fujimoto, W., Robertson, R.P. (1983) Lipoxygenase pathway in islet endocrine cells. J. Clin. Invest. 71: 1191-1205

366 Michelson, A.M., Puget, K. (1980) Cell penetration by exogenous superoxide dismutase. Acta Physiol. Scand. Suppl. 492: 67-80

367 Minchin, R.F., Boyd, M.R. (1983) Localization of metabolic activation and deactivation system in the lung: significance to the pulmonary toxicity of xenobiotics Annu Rev. Pharmacol. Toxicol. 23: 217-238

368 Mittal, C.K., Murad, F. (1977) Activation of guanylate cyclase by superoxide dismutase and hydroxyl radical: a physiological regulator of guanosine 3',5'-monophosphate formation. Proc. Natl. Acad. Sci. USA 74: 4360-4364

369 Mizuno, Y. (1984) Changes in superoxide dismutase, catalase, glutathione peroxidase and glutathione reductase activies and thiobarbituric acid-reactive products levels in early stages of development in dystrophic chickens. Exp. Neurol. 84: 58-73

370 Mizuno, Y. (1984) Superoxide dismutase activity in early stages of development innormal and dystrophic chickens Life Sci. 34: 909-914

371 Mohanam, S., Bose, S.M. (1983) Influence of streptozotocin- and alloxan-induced diabees in the rat on collagenase and certain lysosomal enzymes, relation to the degradation on connective tissue protein. Diabetologia 25: 60-70
372 Monboisse, J.C., Braquet, P., Randoux, A., Borel, J.P. (1983) Non-enzymatic degradation of acid-soluble calf skin collagen by superoxide ion: protective effect of flavonoids. Biochem. Pharmacol 32: 53-58
373 Moncada, S., Vane, J.R. (1981) Prostacyclin and blood coagulation. Drugs 21: 430-437
374 Moncada, S. (1979) Prostacyclin and thromboxane A_2 - platelet vessel wall interactions. In: Gotto AM Ir, Smith LC, Allen B (eds) Atherosclerosis V. Proceedings of the fifth international symposium. Springer, Berlin Heidelberg New York
375 Moneret-Vautrin, D.A., Martin, R. (1985) Intolerance to aspirin: a role for free radicals? Lancet i: 929
376 Montsko, T., Tigyi, A., Zsoldos, T. (1984) Morphological and histochemical study of membrane damages occurring in the early phase of experimental silicosis. In: Oxygen free radicals and tissue damage, scientific session, Pécs, 10-11 January 1984
377 Moody, C.S., Hassan, H.M. (1982) Mutagenicity of oxygen free radicals. Proc. Natl. Acad. Sci. USA 79: 2855-2859
378 Morel, D.W., DiCorleto, P.E., Chisom, G.M. (1984) Endothelial and smooth muscle cells alter low density lipoprotein in vitro by free radical oxidation Arteriosclerosis 4: 357-364
379 Moreno, S.N.J., Mason, R.P., Docampo, R. (1984) Distinct reduction of nitrofurans and Metronidazole to free radical metabolites by Tritrichomonas foetus hydrogenosomal and cytosolic enzymes. J. Biol. Chem. 259: 8252-8259
380 Moreno, S.N.J., Mason, R.P., Docampo, R. (1984) Reduction of Nifurtimox and Nitrofurantoin to free radical metabolites by rat liver mitochondria. J. Biol. Chem. 259: 6298-6305
381 Morisaki, N., Lindsey, J.A., Stitts, J.M., Zhang, H., Cornwell, D.G. (1984) Fatty acid metabolism and cell proliferation V. Evaluation of pathways for the generation of lipid peroxides. Lipids 19: 381-394
382 Mossmann, B., Light, W., Wei, E. (1983) Asbestos: mechanisms of toxicity and carcinogenicity in the respiratory tract. Annu Rev. Pharmacol. Toxicol. 23: 595-614
383 Mountz, J.D., Downs Minor, M.B., Turner, R., Tohams, M.B., Richaros, F., Pisko, E. (1983) Bleomycin-induced cutaneous toxicity in the rat: analysis for histophathology and ultrastructure compared with progressive systemic sclerosis (scleroderma). Br J. Dermatol. 108: 679-686
384 Mózsik, Gy., Jávor, T., Tigyi, A., Tóth, Gy., Zsoldos, T. (1983) Interrelationships between the development of ethanol: HCl, NaCl, NaOH-induced gastric mucosal damage cytoprotective effects of vitamin A and betacarotene, and gastric mucosal superoxide dismutase activity in rats. In: International symposium on the recent advances in gastrointestinal cytoprotection, Pécs, September 30-October 1, 1983
385 Mukai, F.H., Goldstein, B.D.: Mutagenicity of malonaldehyde, decomposition product of peroxidized polyunsaturated fatty acids. Science 191: 868-869
386 Muller, D.P.R., Lloyd, J.K. (1982) Effect of large oral doses of Vitamin E on the neurological sequeleae of patients with abetalipoprotenaemia. In: Lubin, B., Machlin, L.J. (eds) Vitamin E: biochemical, hematological, and clinical aspects. Ann. NY Acad. Sci. 393: 133-142

387 Mustard, J.F., Packham, M.A. (1984) Platelets and diabetes mellitus. N. Engl. J. Med. 311: 665-667
388 Muto, T., Tanaka, J., Miura, T., Kimura, M. (1983) Autooxidation of cholesterol in the presence of unsaturated fatty acid and iron catalyst. J. Pharmacobiodyn. 6: 3-15
389 Müller, A., Sies, H. (1984) Assay of ethane and pentane from isolated organs and cells. Methods Enzymol 105: 311-319
390 Myers, C.E., Katki, A., Trauis, E. (1982) Effect of toxopherol and selenium on defenses against reactive oxygen species and their effect on radiation sensitivity. In: Lubin, B., Machlin, L.J. (eds) Vitamin E: biochemical, hematological, and clinical aspects. Ann. NY Acad.Sci. 393: 419-242
391 Myers, L.S. (1973) Free radical damage of nucleic acids and their components by ionizing radiation. Fed. Proc. 32: 1882-1894.
392 Nabi, Z.F., Rabinovitch, M. (1984) Inhibition by superoxide dismutase and catalase of the damage of isolated Leishmania Mexicana Amazoniensis by phenazine methosulfate. Mol. Biochem. Parasitol 10: 297-303
393 Nagasawa, H., Little, J.B. (1981) Factors influencing the induction of sister chromatid exchanges in mammalian cells by 12-O-tetradecanoyl-phorbol-13-acetate. Carcinogenesis 2: 601-607
394 Nakagawara, A., De Santis, N.M., Nogreira, N., Nathan, C.F. (1982) Lymphokines enhance the capacity of human monocytes to secrete reactive oxygen intermediates. J. Clin. Invest. 70: 1042-1048
395 Nakayama, T., Kodama, M., Nagata, C. (1984) Free radical formation in DNA by lipid peroxidation. Agric Biol Chem 48: 571-572
396 Nath, N., Chari, S.N., Rathi, A.B. (1984) Superoxide dismutase in diabetic polymorphonuclear leukocytes. Diabetes 33: 586-589
397 Nathan, C.F. (1983) Mechanisms of macrophage antimicrobial activity. Trans R Soc. Trop. Med. Hyg 77: 620-630
398 Nera, E.A., Lok, E., Iverson, F., Ormsby, E., Karpinski, K.F., Clayson, D.B. (1984) Short term pathological and proliferative effects of butylated hydroxianisole and other phenolic antioxidants in the forestomach of Fischer 344 rats. Toxicology 32: 197-213
399 Néve, J., Sinet, P.M., Molle, L., Nicole, A. (1983) Selenium, zinc and copper in Down's syndrome (trisomy 21): blood levels and relations with glutathione peroxidase and superoxide dismutase. Clin. Chim. Acta 133: 209-214
400 Niki, E., Saito, T., Kamiya, Y. (1983 The role of vitamin C as an antioxidant. Chem Lett: 631-632
401 Nilsson, B. (1984) Probable in vivo induction of differentiation by retinoic acid of promylocytes in acute promyelocytic leukaemia. Br J. Haematol. 57: 363-371
402 Nissen, H.P., Kreysel, H.W. (1983) Superoxide dismutase in human semen. Klin Wochenschr. 61: 63-65
403 Niwa, Y., Sakane, T., Miyachi, Y., Ozaki, M. (1984 Oxygen metabolism in phagocytes of leprotic patients: enhanced endogenous superoxide dismutase activity and hydroxyl radical generation by Clofazimine. J. Clin. Microbiol. 20: 837-842
404 Nohl, H. (1981) Physiologische und pathophysiologische Bedeutung von Superoxid-Radikalen und die regulatorische Rolle des Enzyms Superoxiddismutase. Klin. Wochenschr. 59: 1081-1091
405 Nohl, H., Hegner, D. (1978) Do mitochondria produce oxygen radicals in vivo? Eur. J. Biochem. 82: 563-567
406 Novelli, G.P. (1985) A spin-trapper protects rats against otherwise lethal shock.

In: Fourth international conference on superoxide and superoxide dismutase, Rome, 1-6 September 1985, abstract p 176

407 Oberley, C.W., Buettner, G.R. (1979) Role of superoxide dismutase in cancer: a review. Cancer Res. 39: 151-159

408 Oberley, L.W. (1984) The role of superoxide dismutase and gene amplification in carcinogenesis. J. Theor. Biol. 106: 403-422

409 Odline, B., Bayati, A., Frödin, L., Heilberg, O., Rutili, F., Bylund AD, Wolgast, M. (1985) Prevention of acute renal failure by oxygen free radical scavengers. In: Fourth international conference on superoxide and superoxide dismutase, Rome, 1-6 September 1985, abstract p 80

410 Oei, H.H., Bose, S.K., McCord, J.M. (1985) Ethanol-induced oxidative stress: role of acetaldehyde in determining xanthine dehydrogenase/oxidase ratio. In: Fourth international conference on superoxide and superoxide dismutase, Rome, 1-6 September 1985, abstract p 89

411 Örndahl, G., Rinby, A., Selin, E. (1983) Selenium therapy of myotonic dystrophy. Acta Med. Scand. 213: 237-239

412 Örndahl, G., Rinby, A., Selin, E. (1982) Myotonic dystrophy and selenium. Acta Med. Scand. 211: 493-499

413 Oroszlán, Gy., Szabó, L., Lakatos, I., Matkovics, B. and Karmazsin, L. (1982) D-penicillinamine doesn't reduce heme oxygenase activity of the newborn. (in Hungarian) Kisérl. Orvostud. 34: 446-448

414 Ovesen, L. (1984) Vitamin therapy in the absence of obvious deficiency. Drugs 27: 148-170

415 Paller, M.S., Heidal, J.R., Ferris, T.F. (1984) Oxygen free radicals in ischemic acute renal failure in the rat. J. Clin. Invest. 74: 1156-1164

416 Panagamala, R.V., Cornwell, D.G. (1982) The effects of Vitamin E on arachidonic acid metabolism. Ann. NY Acad. Sci. 393: 376-390

417 Paniker, N.V., Arnold, A.B., Hartmann, R.C. (1974) Studies of the role of red cell membrane peroxidation in paroxysmal nocturnal haemoglobunuria (PNH). Br. J. Haematol. 26: 39-47

418 Pár, A., Jávor, T. (1984) Alternatives in hepatoprotection: cytoprotection-influences on monooxidase system-free radical scavengers. Acta Physiol. Hung. 64: 409-423

419 Parks, D.A., Bulkley, G.B., Granger, D.N. (1983) Role of oxygen free radicals in shock, ischaemia and organ preservation. Surgery 94: 428-432

420 Parks, D.A., Bulkley, G.B., Granger, D.N. (1983) Role of oxygen-derived free radicals in digestive tract disease. Surgery 94: 415-422

421 Parks, D.A., Granger, O.N., Bulkley, G.B. (1982) Superoxide radicals and mucosal lesions in the ischaemic small intestine. Fed. Proc. 141: 1742

422 Partridge, S.M. (1969) Elastin, biosynthesis and structure. Gerontologia 15: 85

423 Pasquier, C., Mach, P.S., Raichvarg, D., Sarfati, G., Amor, B., Delbarre F. (1984) Manganese-containing superoxide dismutase deficiency in polymorphonuclear leukocytes of adults with rheumatoid arthritis. Inflammation 8: 27-32

424 Pathak, M.A., Fitzpatrick, T.B., Parrish, J.A. (1980) Photosensitivity and other reactions to light. In: Harrison's principles of internal medicine, q + h edn. McGraw Hill, Kokagusha, pp 255-262

424a Pár, A. (1980) Levamisole as an immunostimulant. Ther Hung 28: 163-167

425 Pár, A., Horváth, T., Pakodi, F., Zsoldos, T., Kerekes, E., Paál, M., Kádal, I., Beró, T., Jávor, T. (1984) Hepatoprotective and immune moderator effects of antioxidants. In: Oxygen free radicals and tissue damage, scientific session, Pécs, 10-11 January 1984

426 Pearce, M.L., Dayton, S. (1971) Incidence of cancer in man on a diet high in polyunsaturated fat Lancet i: 464-467
427 Peck, G.D. (1982) Retinoids. Therapeutic use in dermatology. Drugs 24: 341-351
428 Pénzes, L., Boross, M., Gergely, I. (1984) Causes of aging, and biochemical possibilities for retardation of aging process. In: Beregi E (ed) The aging process (in Hungarian). Academic, Budapest, pp 18-38
429 Pesanti, E.L. (1984) Pneumocystis carinii: oxygen uptake, antioxidant enzymes and susceptibility to oxygen-mediated damage. Infect. Immun. 44: 7-11
430 Peters, T.J., Seymour, C.A. (1979) The organelle pathology and demonstration of mitochondrial superoxide dismutase deficiency in two patients with Dubin-Johnson-Sprinz syndrome. Clin. Sci Mol Medicine 54: 549-553
431 Petkai, A. (1980) Radiation carcinogenesis from a membrane perspective. In: Free radicals - medicine and biology. Acta Physiol. Scand. Suppl. 492: 81-90
432 Petrone, W.F., English, O.K., Wong, K., McCord, J. (1980) Free radicals and inflammation: superoxide-dependent activation of a neutrophil chemotactic factor in plasma. Proc. Natl. Acad. Sci. USA 77: 1159-1163
433 Petrovic, V.M., Spasic, M., Saicic, Z., Milic, R. and Radijicic, R. (1982) Increase in superoxide dismutase activity induced by thyroid hormones in the brains of neonate and adult rats. Experientia 38: 1355-1356
434 Pine, L., Hoffman, P.S., Malcolm, G.B., Benson, R.F., Keen, M.G. (1984) Determination of catalase, peroxidase and superoxide dismutase within the genus *Legionella*. J. Clin. Microbiol. 20: 421-429
435 Plaa, G.L., Witschi, H. (1976) Chemicals, drugs and lipid peroxidation. Annu Rev. Pharmacol. Toxicol. 161: 125-141
436 Player, T.J., Mills, D.J., Horton, A.A. (1979) Lipid peroxidation of the microsomal fraction and extracted microsomal lipids from DAB-induced hepatomas. Br. J. Cancer 39: 773-778
437 Pollakis, G., Goormaghtigh, E., Delmelle, M., Lion, Y, Ruysschaert, J-M. (1984) Adriamycin and derivates interaction with the mitochondrial membrane: O_2 consumption and free radicals formation. Res. Commun. Chem. Pathol. Pharmacol. 44: 445-459
438 Pollakis, G., Goormaghtigh, E., Ruysschaert, J-M. (1983) Role of the quinone structure in the mitochondrial damage induced by antitumor anthracyclines. FEBS Lett 155: 267-272
439 Pottathil, R., Chandrabose, K.A., Chatrecasas, P., Lang, D.J. (1981) Establishment of the interferon mediated antiviral state: possible role of superoxide dismutase. Proc. Natl. Acad. Sci. USA 78: 3343-3347
440 Porter, N.A. (1980) Prostaglandin endoperoxides. In: Pryor, W.A. (ed) Free radicals in biology, vol VI. Academic, New York, pp 261-294
441 Prosbkina, M.P., Matyuskin, A.I. (1982) Effect of hormones on lipid peroxidation in the heart and liver. Farmakol toxikol 45: 24-26
442 Pryor, W.A. (1982) Free radical biology: xenobiotics, cancer and aging. In: Lubin B, Machlin LT (eds) Vitamin E: biochemical, hematological, and clinical aspects. Ann. NY Acad. Sci. 393: 1-22
443 Pryor, W.A. (1973) Free radical reactions and their importance in biochemical systems. Fed. Proc. 32 1862-1869
444 Pryor, W.A., Castle, L. (1984) Chemical methods for the detection of lipid hydroperoxides. Methods Enzymol 105: 293-299
445 Pryor, W.A., Prier, D.G., Church, D.F. (1983) Electron spin resonance study of mainstream and sidestream cigarette smoke: nature of the free radicals in gas-phase smoke and in cigarette tar. Environ Health Perspect 47: 345-355

446 Pryor, W.A. (1982) Free radical biology: xenobiotics, cancer and aging. Ann. NY Acad. Sci. 293: 1-22
447 Rachmilewitz, E.A., Kornberg, A., Acker, M. (1982) Vitamin E deficiency due to increased consumption in beta-thalassaemia and in Gaucher's disease. In: Lubin, B., Machlin, L.J. (eds) Vitamin E: biochemical, hematological, and clinical aspects. Ann. NY Acad. Sci. 393: 336-346
448 Raj, A.S., Heddle, J.A. (1980) The effect of superoxide dismutase, catalase and L-cysteine on spontaneous and on mitomycin C induced chromosomal breakage in Fanconi's anaemia and normal fibroblasts as measured by the micronucleus method. Mutat. Res. 78: 59-66
449 Ramos, K., Combs, A.B., Acosta, D.: Cytotoxicity of isoproterenol to cultured heart cells: effects of antioxidants on modifying membrane damage. Toxicol. Appl. Pharmacol. 70: 317-323
450 Rao, M.S., Laiwani, N.D., Watanabe, T.K., Reddy, J.K. (1984) Inhibitory effect of antioxidants Ethoxyquin and 2(3)-tert-butyl-4-hydroxyanisole on hepatic tumorigenesis in rats fed Ciprofibrate a peroxisome proliferator. Cancer Res. 44: 1072-1076
451 Realini, S., Gonvers, J.J., Hofstetter, J.R. (1975) Essai clinique de la silymarine dans les affections chroniques du foie. Schweiz. Rdsch. Med. 64: 595
452 Reimer, K.A., Jennings, R.B., Tatum, A.H. (1983) Pathology of acute myocardial ischaemia: metabolic, functional and ultrastructural studies. Am J. Cardiol. 52: 72A/81A
453 Reizenstein, P., Máthé, G.: Biological response modifiers and differentiation inducers. Drugs 26 185-190
454 Renner, H.W. (1984) Antimutagenic effect of an antioxidant in mammals. Mutat Res. 135: 125-129
455 Reynolds, E.S., Treinen, S., Moslen, H. (1980) Free-radical damage in liver. In: Pryor, W.A. (ed) Free radicals in biology, vol. IV. Academic, New York, pp 49-94
456 Ribarov, S.R., Benov, L.C., Benchev, I.C. (1984) $HgCl_2$ increases the methemoglobin prooxidant activity. Possible mechanism of Hg^{2+}-induced lipid peroxidation in erythrocytes. Chem. Biol. Interact 50: 11-119
457 Ribiére, C., Sinaceur, S., Nordmann, J., Nordmann, R. (1983) Liver superoxide dismutases and catalase during ethanol inhalation and withdrawal. Pharmacol. Biochem. Behav. 18, suppl. 1: 263-266
458 Rister, H., Bauermeister, K., Grauert, V., Gladtke, E. (1978) Superoxide-dismutase deficiency in rheumatoid-arthritis. Lancet i: 1094
459 Rister, M., Bauermeister, K. (1982) Superoxid-dismutase und Superoxid-Radikal-Freisetzung bei juweniler rheumatoider Arthritis. Klin. Wochenschr. 60: 561-565
460 Robert, J.A., Roth, K.J., Domingue, F., Lewis, R.W., Kaack, B., Baskin, G. (1982) Immunology of pyelonephritis in the primate model V. Effect of superoxide dismutase. J. Urol. 128: 1394-1400
461 Robertson, M. (1983) Oncogenes and the origins of human cancer. Br. Med. J. 286 (6359): 181-182
462 Robertson, M. (1983) Oncogenes and multistep carcinogenesis. Br. Med. J. 287: 1084-1086
463 Robertson, R.P. (1983) Prostaglandin glucose homeostasis and diabetes mellitus. Annu Rev. Med. 34: 1-12
464 Roisman, F.R., Walz, D.T., Finkelstein, A.E. (1983) Superoxide radical production by human leukocytes exposed to immune complexes: inhibitory action of gold compounds. Inflammation 7: 355-362

465 Romson, J.L., Hodk, G.B., Kunkel, S.L., Abrams, G.D., Schork, K.M.A., Lucchesi, B.R. (1983) Reduction of the extent of ischemic myocardial injury by neutrophil depletion in the dog. Circulation 67: 1016-1023

466 Roos, D., Bragt, P.C. (1984) Workshop session summary: Is phagocyte radical production relevant to inflammation? Agents Actions 15: 57-58

467 Root, R.K., Metcalf, J.A. (1977) H_2O_2 release from human granulocytes during phagocytosis. J. Clin. Invest. 60 1266-1279

468 Rosenfeld, W., Evans, H., Concepcion, L., Jhaveri, R., Schaeffer, H., Friedman, A. (1984) Prevention of bronchopulmonary dysplasia by administration of bovine superoxide dismutase in preterm infants with respiratory distress syndrome. J. Pediatr. 105: 781-785

469 Rowe, G.T., Eaton, L.R., Hess, M.L. (1984) Neutrophil-derived, oxygen free radicalmediated cardiovascular dysfunction. J. Mol. Cell. Cardiol. 16: 1075-1079

470 Rowley, D.A., Halliwell, B. (1983) Formation of hydroxyl radicals from hydrogen peroxide and iron salts by superoxide and ascorbate-dependent mechanism: relevance to the pathology of rheumatoid disease. Clin. Sci. 64: 649-653

471 Röth, E., Török, B., Zsoldos, T., Matkovics, B., Szabó, L. (1984) Lipid peroxidation in experimental myocardial infarction. In: Oxygen free radicals and tissue damage, scientific session, Pécs, 10-11 January 1984

472 Rubach, L., Warter, J.M., Coquillat, G., Marescaux, C., Collard, M., Rohmer, F., Bieth, R., Zawislak, R. (1983) Anomalies de la répartition des acides gras et de l'activité de la superoxide dismutase dans les lymphocytes d'un adulte atteint de céroide-lipofuscinose atypique. Rev. Neurol. (Paris) 139: 269-276

473 Ruprah, M., Mant, T.G.K., Flanagan, R.J. (1985) Acute carbon tetrachloride poisoning in 19 patients: implications for diagnosis and treatment. Lancet i: 1027-1029

474 Ryle, P.R. (1984) Free radicals, lipid peroxidation and ethanol hepatotoxicity. Lancet ii: 461

475 Sacks, T., Moldow, C.F., Craddock, P.R., Bowers, T.K., Jacob, H.S. (1978) Oxygen radicals mediate endothelial cell damage by complement stimulated granulocytes. J. Clin. Invest. 61: 1161-1167

476 Sagai, M., Ichinose, T., Kubota, K. (1984) Studies on the biochemical effects of nitrogen dioxide. Toxicol. Appl. Pharmacol. 73: 444-456

477 Sagone, A.L.Jr, Greenwald, J., Kraut, F., Bianchine, J., Singh, D. (1983) Glucose: a role as a free radical scavenger in biological systems. J. Lab. Clin. Med. 101: 97-104

478 Salin, M.L., McCord, J.M. (1975) Free radicals and inflammation. J. Clin. Invest. 56: 1319-1223

479 Samuelsson, B. (1982) Die Leukotriene superaktive an Allergie und Entzündung beteiligte Wirkstoffe. Angew Chem 94: 881-982

480 Sanfey, H., Bulkley, G.B., Cameron, J.L. (1984) The role of oxygen derived free radicals in the pathogenesis of acute pancreatitis. Ann. Surg. 200: 405-413

481 Saul, R.L., Ames, B.N. (1985) Oxidative DNA damage as related to cancer and aging. In: 1st international congress of Biomedical Gerontology, New York, 10-11 July 1985, abstract 52

482 Sawas, A.H., Gilbert, J.C. (1984) Possible mechanism of inhibition by lipid peroxidation of ATP-ase activities of rat cerebral cortex synaptosomes. Arch. Int. Pharmacodyn. Ther. 269: 4-11

483 Schersten, T. (1985) The role of oxygen free radicals for postischemic kidney

damage. In: Fourth international conference on superoxide and superoxide dismutase, Rome, 1-6 September 1985, abstract L 8
484 Schneider, E.L., Reed, J.D. (1985) Life extension. N. Engl. J. Med. 312: 1159-1168
485 Schwarz, M., Peres, G., Kunz, W., Fürstenberger, G., Kittstein, W., Marks, F. (1984) On the role of superoxide anion radicals in skin tumour promotion. Carcinogenesis 5: 1663-1670
486 Sealy, R.C., Felix, C.C., Hyde, J.S., Schwartz, H.M. (1980) Structure and reactivity of melanins: influence of free radicals and metal ions. In: Pryor, W.A. (ed) Free radicals in biology, vol. IV. Academic, New York, pp. 261-294
487 Segal, A.W., Cross, A.R., Garcia, R.C., Borregaard, N., Valerius, N.H., Soothill, J.F., Jones, O.T.G. (1983) Absence of cytochrome b-245 in chronic granulomatous disease A multicenter European evaluation of its incidence and relevance. N. Engl. J. Med. 308: 245-251
488 Sevanian, A. (1985) Free-radical induced damage to membrane phospholipids and their susceptibility to phospholipase A_2. In: Fourth international conference on superoxide and superoxide dismutase, Rome, 1-6 September, 1985, abstract p 75
489 Shah, S.V., Wallin, J.D., Eilen, S.D. (1983) Chemiluminescence and superoxide anion production by leukocytes from diabetic patients. J. Clin. Endocrinol. Metab. 57: 402-409
490 Shamberger, R.J. (1974) Antioxidants and cancer IV. Initiating activity of malonaldehyde as a carcinogen. J. Natl. Cancer Inst. 53: 1171-1773
491 Sharma, S.P., Wadhwa, R. (1985) Effect of sodium hypophosphite on longevity and nucleic acid content in aging Zaprionus Paravittiger. In: 1st international congress of biomedical Gerontology, New York, 10-11 July 1985, abstract 67
492 Shaw, S., Jayatilleke, E., Ross, W.A., Gordon, E.F., Lieber, C.S. (1981) Ethanol-induced lipid peroxidation potentiation by long-term alcohol feeding and attenuation by methionine. J. Lab. Clin. Med. 98: 417-424
493 Shaw, S., Rubin, K.P., Lieber, C.S. (1983) Depressed Hepatic glutathione and increased diene conjugates in alkoholic liver disease. Evidence of lipid peroxidation. Digest. Dis. Sci. 28: 585-589
494 Shlafer, M., Kane, P.F., Valerie, Y., Wiggins, B.S., Kirsch, M.M. (1982) Possible role for cytotoxic oxygen metabolites in the pathogenesis of cardial ischemic injury. Circulation 66, suppl 1: I85-I92
495 Shohet, S.B.: Haemolysis and changes in erythrocyte membrane lipids (second of two parts). N. Engl. J. Med. 286: 638-644
496 Shohet, S.B., Jain, S.K. (1982) Introduction: Vitamin E and blood cell function. In: Lubin, B., Machlin, L.J. (eds) Vitamin E: biochemical, hematological, and clinical aspects. Ann. NY Acad. Sci. 393: 229-236, 1982
497 Siakotos, A.N., Koppang, D.V.M., Youmans, B.S., Bucana, C. (1974) Blood levels of alpha-tocopherol in a disorder of lipid peroxidation: Batten's disease. Am J. Clin. Nutr. 27: 1152-1157
498 Simchowitz, L., Spilberg, I. (1979) Evidence for the role of superoxide radicals in neutrophil-mediated cytotoxicity. Immunology 37: 301-309
499 Singal, P.K., Kapur, N., Dhillon, K.S., Beamish, R.E., Dhalla, N.S. (1982) Role of free radicals in catecholamine-induced cardiomyopathy. In: Active oxygen and medicine symposium. Can J. Physiol. Pharmacol. 60: 1390-1397
500 Singh, A. (1982) Chemical and biochemical aspects of superoxide radicals and related species of activated oxygen. In: Active oxygen and medicine symposium. Can J. Physiol. Pharmacol. 60: 1330-1345

501 Sinha, B.K. (1984) Metabolic activation of procarbazine Evidence for carbon-centered free radical intermediates. Biochem. Pharmacol 33: 2777-2781

502 Siriwittayakorn, J., Yuthavong, Y. (1979) Relation between low erythrocyte acetyl-cholesterase activity and membrane lipids in paroxysmal nocturnal haemoglobinuria. Br. J. Haematol. 41: 383-391

503 Slater, T.F. (1984) Review article: Free-radical mechanisms of tissue injury Biochem. J. 222: 1-15

504 Slater, T.F. (1981) Free radical scavengers. In: Conn HO (ed) International workshop on (+)-cyanidanol-3 in disease of the liver. pp 11-15 (Royal Society of Medicine International Congress and Symposium Series no 47

505 Slater, T.F., Scott, R. (1981) The free radical scavenging action of (+)-cyanidanol-3 in relation to the toxicity of carbon tetrachloride. In: Conn HO (ed) International Workshop on (+)-3-cyanidanol-3 in diseases of the liver, pp 33-39 (Royal Society of Medicine International Congress and Symposium Series no 47)

506 Smith, D.S., Renchrona, S., Siesjö, B.K. (1980) Barbiturates as protective agents in brain ischaemia and as free radical scavengers in vitro. In: Free radicals in medicine and biology symposium. Acta Physiol. Scand. Suppl. 492: 129-134

507 Smith, R.C., Lawing, L. (1983) Antioxidant activity of uric acid and 3-N-ribosyluric acid with unsaturated fatty acids and erythrocyte membranes. Arch. Biochem. Biophys. 223: 166-172

508 Smolen, J.E. (1984) Lag period for superoxide anion generation and lysosomal enzyme release from human neutrophils effects of calcium antagonists and anion channel blockers. J. Lab. Clin. Med. 104: 1-10

509 Snyder, L.M., Fortier, N.L., Sauberman, N., Leb, L., Trainor (Piotrowski), J., Fairbanks, G. (1984) Hemoglobin-spectrin interaction under oxidative conditions. In: Recent clinical and experimental advances, Liss, New York, pp 227-242 (Erythrocyte membranes, vol 3)

510 Sofuni, T., Ishidate, M.Jr. (1984) Induction of chromosomal aberrations in cultured Chinese hamster cells in a superoxide-generating system. Mutat Res 140: 27-31

511 Sohal, R.S., Allen, R.G., Farmer, K.J., Proctor, J. (1984) Effect of physical activity on superoxide dismutase, catalase, inorganic peroxides and glutathione in the adult male housefly, Musca Domestica. Mech. Ageing Dev 26: 75-81

512 Sohal, R.S., Farmer, K.J., Allen, R.G., Cohen, N.R. (1983) Effect of age on oxygen consumption, superoxide dismutase, catalase, glutathione, inorganic peroxides and chloroform-soluble antioxidants in the adult male housefly, Musca Domestica. Mech. Ageing Dev 24: 185-195

513 Solar, C.J. (1959) Vitamin E and ulcers. Rev. Esp Enferm Apar Dig Nutr. 18: 745

514 Sorensen, J.R.J. (1984) Bis (3,5-diisopropylsalicylate)copper(II), a potent radioprotectant with superoxide dismutase mimetic activity. J. Med. Chem. 27: 1747-1749

515 Spisani, S., Marangoni, C., Traniello, S. (1984) Effect of anti-inflammatory drugs on leukocyte superoxide production by soluble and particulate stimuli. Biochem. Int. 9: 1-8

516 Sréter, L., Toncsev, H., Fekete, S., Romics, L., Fehér, J., Gráf, F. (1983) Investigation of serum and granulocyte beta-glucuronidase activity in cases of joint occurrence of hyperlipidaemia and hepatic steatosis. Kisérl. Orvostud. 35: 583-586

517 Sréter, L., Kiss, A., Cornides, A., Vereckei, A., Toncsev, H., Fehér, J. (1984)

Inhibiton of Doxorubicin-induced hepatic toxicity by new dihydroquinoline type antioxidant. In: International symposium on recent advances in gastrointestinal cytoprotection, Pécs, 30 September-1 Oktober 1983. Acta Physiol. Acad. Sci. Hung. 64: 431-435

518 Srivastava, V.K., Singh, J.P. Srivastava, S.K. (1985) On the formation of senile cataract-implications of superoxide radicals (O_2^-). Fourth international conference on superoxide and superoxide dismutase, Rome, 1-6 September 1985, abstract p 76

519 Stacey, N.H., Kappus, H. (1982) Cellular toxicity and lipid peroxidation in response to mercury. Toxicol. Appl. Pharmacol. 63: 29-35

520 Stacey, N.H., Klaassen, C.D. (1982) Effects of phospholipase A_2 inhibitors on diethyl maleate-induced lipid peroxidation and cellular injury in isolated rat hepatocytes. J. Toxicol. Environ. Health 9: 439-450

521 Starkebaum, G., Dennis, L., Stevens, C.L., Gavin, S.E. (1981) Stimulation of human neutrophil chemiluminescence by soluble immune complexes and antibodies to neutrophils. J. Lab. Clin. Med. 98 280-291

522 Stege, T.E., Mischke, B.S., Cox, G.W., Daniels, K.A. (1983) The role of free-radical inhibitors on acetaldehyde induced increases in lipid peroxidation. Fed. Proc. 47: 513

523 Stenmark, K.R., James, S.L., Voelkel, N.F., Toews, W.H., Reeves, J.T., Murphy, R.C. (1983) Leukotriéne C_4 and D_4 in neonates with hypoxaemia and pulmonary hypertension. N. Eng. J. Med. 309: 77-80

524 Stern, L.Z., Ringel, S.P., Ziter, F.A., Menander-Huber, K.B., Ionasescu, U., Pellegrino, R.J., Snyder, E.D. (1982) Drug trial of superoxide dismutase in Duchenne's muscular dystrophy. Arch. Neurol. 39: 342-346

525 Stickle, D.F., Daniele, R.P., Holian, A. (1984) Cytosolic calcium, calcium fluxes and regulation of alveolar macrophage superoxide anion production. J. Cell. Physiol 121: 458-466

526 Sugioka, K., Nakano, M. (1982) Mechanism of phospholipid peroxidation induced by ferric ion-ADP-adriamycin co-ordination complex. Biochim. Biophys. Acta 713: 333-342

527 Sundaram, G.G., Manmekalai, S., London, R.S., Goldstein P (1982) Effect of alpha-tocopherol on serum hormone and lipoprotein cholesterol levels in young and adult female rats. In: Lubin, B., B, Machlin, L.J. (eds) Vitamin E: biochemical, hematological and clinical aspects. Ann. NY Acad. Sci. 393: 190-192

528 Sung, C., Mirabelli, C.K., Badger, A.M. (1984) Effect of gold compounds on phorbol myristic acetate (PMA) activated superoxide (O_2^-) production by mouse peritoneal macrophages. J. Rheumatol. 11: 153-157

529 Suttorp, N., Simon, L.M. (1982) Lung cell oxidant injury enhancement of polymorphonuclear leukocyte-mediated cytotoxicity in lung exposed to sustained in vitro hyperoxia. J. Clin. Invest. 70: 342-350

530 Sülle, Cs., Vladimirovich, M., Konstantinovich, Zh. A. (1985) Chemoluminescence in gastric and duodenal ulcer and in chronic and erosive gastritis (in Hungarian). Hung. Arch. Int. Med. 38: 5-13

531 Szabó, L.Sz., Varga, I., Matkovics, B. (1984) Methods of determination of lipid peroxidation. In: Oxygen free radicals and tissue damage. Scientific session, Pécs, 10-11 January 1984

532 Szajfutgyinov, R.G. (1982) The effect of epinephrine on the level of free radicals in human plasma and erythrocytes (in Russian). Biochim Biphys: 78-79

533 Takahashi, M., Yoneda, Y., Kuriyama, K. (1984) Enhancement of gamma-aminobutiric acid (GABA) receptor binding by lipophilic antioxidants. Brain Res. 296: 164-167

534 Tanimura, H., Kato, H., Hilkasa, Y. (1982) The role of vitamin E in the etiology and treatment of pancreatitis. In: Lubin, B., Machlin L.J. (eds) Vitamin E: biochemical, hematological, and clinical aspects. Ann. NY Acad. Sci. 393: 214-216

535 Tappel, A.L. (1973) Lipid peroxidation damage to cell components. Fed. Proc. 32: 1870-1874

536 Tappel, A.L., Dillard, C.U. (1981) In vivo lipid peroxidation: measurement via exhaled pentane and protection by vitamin E. Physiological significance of lipid peroxidation. Fed. Proc. 40: 174-178

537 Tate, R.M., Repine, J.E. (1984) Phagocytes, oxygen radicals and lung injury. In: Pryor WA (ed) Free radicals in biology Academic, New York, pp 199-209

538 Thaete, L.G., Crouch, R.K., Schulte, B.A., Spicer, S.S. (1983) The immunolocalization of copper-zinc superoxide dismutase .in canine tissues. J. Histochem. Cytochem. 31: 1399-1406

539 Theelen, M., Wendel, A. (1983) Drug induced lipid peroxidation in mice-V Ethane production and glutathione release in the isolated liver upon perfusion with acetaminophen. Biochem. Pharmacol 32: 1701-1706

540 Tigyi, A., Zsoldos, T., Monstkó, T. (1984) Pathogenesis of experimental silicosis. In: Oxygen free radicals and tissue damage, scientific session, Pécs, 10-11 January 1984

541 Tisdale, M.J., Mahmoud, M.B. (1983) Activities of free radical metabolizing enzymes in tumors. Br. J. Cancer 47: 809-812

542 Toibana, E., Ohnishi, R., Fujimoto, Y., Fujita, T. (1983) Effect of mercurials on lipid peroxidation in rabbit renal cortical mitochondria. Jap. J. Pharmacol. 33: 1279-1281

543 Toncsev, H., Fehér, J., Fehér, E. (1980) Activity and release of beta-glucuronidase in granulocytes of patients with chronic liver diseases. Hepatology 9: 22-23

544 Toncsev, H., Fehér, J., Fehér, E., Vasadi, Á. (1981) Investigation of lysosomal enzymes in the serum and granulocytes of patients with chronic hepatic diseases. Magy Belorv. Arch. 34: 293-298

545 Toncsev, H., B. Pollák, Zs., Kiss, Á., Cornides, Á., Fehér, J. (1983) Protective effect of a new dihydroquinoline-type antioxidant on lysosomal damage in carbon tetrachloride induced free radical reactions. Kisérl. Orvostud 34: 16-21

546 Toncsev, H., Sulyok, S., Sréter, L., Vereckei, A., Pollák, Zs., Fehér, J., (1983) Prevention of the damaging effect of cholesterol on the hepatic lysosomal membrane with a dihydroquinoline-type antioxidant. In: Jubilee session of the Hungarain gastroenterological association, Keszthely, 17-20 May 1983

547 Toncsev, H., Fehér, J. (1982) The effect of various lipids on the release from human granulocytes in vitro. In: Treci jugoslovenski simpozyum on hiperlipoprotein-emijama. Novi Sad, 290-294

548 Toncsev, H., Pollák, Zs., Kiss, Á., Sréter, L., Fehér, J. (1982) Acute carbon tetrachloride induced lysosomal membrane damage and the membrane protecting effect of new dihydroquinoline-type antioxidant. Int. J. Tissue React 4: 325-330

549 Totter, J.R. (1980) Spontaneous cancer and its possible relationship to oxygen metabolism. Proc. Natl. Acad. Sci. USA 77: 1763-1767

550 Török, B., Röth, E., Zsoldos, T., Matkovics, B., Szabó, L. (1984) Lipid-membrane interaction in cardiac and pulmonary tissue in dog and rat experiments, in hyperoxia, hypoxia and ischaemia models. In: Oxygen free radicals and tissue damage, scientific session, Pécs, 10-11 January 1984

551 Török, J., Morvay, J., Falkay, Gy. (1981) Lipid peroxidation in the early human placenta. Kisérl. Orvostud. 33: 476-479

552 Triggs, W.J., Willmore, L.J. (1984) In vivo lipid peroxidation in rat brain following intracortical Fe^{2+} injection. J. Neurochem. 42: 976-980
553 Troll, W., Wiesner, R. (1985) The role of oxygen radicals as a possible mechanism of tumor promotion. Annu Rev. Pharmacol. Toxicol. 25: 509-528
554 Trush, M.A., Mimnaugh, E.G., Ginsburg, E., Gram, T.E. (1982) Studies on the interaction of Bleomycin A_2 with rat lung microsomes II. Involvement of adventitious iron and reactive oxygen in Bleomycin mediated DNA chain breakage. J. Pharmacol. Exp. Ther. 221: 159-165
555 Trush, M.A., Mimnaugh, E.G., Gram, T.E. (1982) Activation of pharmacologic agents to radical intermediates. Implications for the role of free radicals in drug action and toxicity. Biochem. Pharmacol. 31: 3335-3346
556 Trush, M.A., Mimnaugh, E.G., Ginsburg, E., Gram, T.E. (1982) Studies on the interaction of Bleomycin-mediated DNA chain breakage. J. Pharmacol. Exp. Ther. 221: 152-158
557 Trush, M.A., Mimnaugh, E.G., Gram, T.E. (1982) Activation of pharmacologic agents to radical intermediates. Implications for the role of free radicals in drug action and toxicity. Biochem. Pharmacol 31: 3335-3346
558 Tseng, Y., Latham, K.R. (1984) Iodothyronines: oxidative diodination by hemoglobin and inhibition of lipid peroxidation Lipids 19: 96-102
559 Turrens, J.F., Crape, J.D., Freeman, B.A. (1984) Protection against oxygen toxicity by intravenous injection of liposoma-entrapped catalase and superoxide dismutase. J. Clin. Invest. 73: 87-95
559a Valenzuela, A., Guorra, R. (1986) Differential effect of silybin on the Fe^{2+}-ADP and t-butyl hydroperoxide induced microsomal lipid peroxidation. Experientia (in press) (to be published)
559b Valenzuela, A., Guorra, R. (1986) Silybin dihemisuccinate protects rat erythrocytes against phenylhydrazine induced lipid peroxidation and hemolysis. Life Sci (to be published)
559c Valenzuela, A., Guorra, R. and Videla, L.A. (1986) Antioxidant properties of the flavonoids silybin and (+)-cyanidanol-3 comparison with butylated hydroxyanisole and butylated hydroxytoluene. Lipids (to be published)
560 Vanella, A., Geremia, E., Pinturo, R., Ririolo, P., Liuzzo, G., Tiriolo, C., Custorella, A., Condorelli, G., Giglio, A. (1983) Superoxide dismutase activity and reduced glutathione content in erythrocytes of uremic patient on chronic dialysis. Acta Haematol 70: 312-315
561 Varma, S.D. (1981) Superoxide and lens of the eye: a new theory of cataractogenesis. Int. J. Quantum Chem 20: 479-484
562 Varma, S.D., Srivastava, U.K., Richards, R.D. (1982) Photoperoxidation in lens and cataract formation: preventive role of superoxide dismutase catalase and vitamin C. Ophthalmic Res. 14 (3): 167-175
563 Vereckei, A., Cornides, Á., Kiss, Á., Sréter, L., Toncsev, H., Fehér, J. (1984) Prevention of doxorubicin-induced liver toxicity by a new dihydroquinoline type antioxidant. Drugs Expt Clin. Res. X(11): 753-757
564 Vistnes, A.I., Henriksen, T., Nicolaissen, B., Armstrong, D. (1983) Free radicals and aging. Electron spin resonance studies on neuronal lipopigments and cells grown in vitro. Mech. Ageing. Develop. 22: 335-345
565 Walford, R.L. (1980) Immunology and aging. Am J. Clin. Path. 74 (9): 247-253
566 Warso, M.A., Lands, W.E.M. (1983) Lipid peroxidation in relation to prostacyclin and thromboxane physiology and pathophysiology. Br Med. Bull. 39: 277-280
567 Warth, J.A., Rucknagel, D.L. (1984) Density ultracentrifugation of sickle cells during and after pain crisis: increased dense echinocytes in crisis. Blood 64: 507-515

568 Watson, B.D., Buste, R., Goldberg, W.J., Santiso, M., Yoshida, S., Ginsberg, M. (1984) Lipid peroxidation in vivo induced by reversible global ischemia in rat brain. J. Neurochem. 42: 268-274
569 Wattenberg, L.W. (1983) Systems detoxifying chemical carcinogens. Cancers Environ: 111-116
570 Wautier, J.L., Paton, R.C., Wautier, M.P., Pintigny, D., Abadie, E., Passa, P., Caen, J.P. (1981) Increased adhesion of erythrocytes to endothelial cells in diabetes mellitus and its relation to vascular complications. N. Engl. J. Med. 305: 237-242
571 Weglicki, W.B., Dickens, B.F., Tong Mak, I. (1984) Enhanced lysosomal phospholipid degradation and lysophospholipid production due to free radicals. Biochem. Biophys. Res. Commun 124: 229-235
572 Weinstein, S.G., Zvershanovsky, F.A. (1984) Effects of the synthetic antioxidants ionol and dibunol on hemorrhagic ulcerative lesions in rat stomach (in Russian). Pharmacol. Toxicol 6: 87-90
573 Weiss, S.J., Lobuglio, A.F. (1982) Phagocyte-generated oxygen metabolites and cellular injury. Lab. Invest. 47: 5-18
574 Weiss, S.J., Turk, S., Needleman, P. (1979) A mechanism for the hydroperoxide-mediated inactivation of prostacyclin synthetase. Blood 53: 1191-1196
575 Weiss, V. (1984) Psychometric intelligence Correlates with interindividual different rates of lipid peroxidation. Biomed. Biochim. Acta 6: 755-763
576 Weissmann, G. (1983) The eicosanoids of asthma. N. Engl. J. Med. 308: 454-456
577 Weissmann, G. (1982) Activation of neutrophils and the lesions of rheumatoid arthritis. J. Lab. Clin. Med 100: 322-333
578 Weiter, J.J. (1981) Retrolental fibroplasia: an unsolved problem. N. Engl. J. Med. 305: 1404-1406
579 Witzmann, S.A., Stossel, T.P. (1981) Mutation caused by human phagocytes. Science 212: 546-547
580 West, P.R., Harman, L.S., Josephy, P.D., Mason, R.P. (1984) Acetaminophen: enzymatic formation of a transient phenoxyl free radical. Biochem. Pharmacol. 33: 2933-2936
581 Willett, W.C., MacMahon, B. (1984) Diet and cancer-an overview (first of two parts). N. Engl. J. Med. 310: 632-638
582 Willett, W.C., MacMahon, B. (1984) Diet and cancer-an overview (second of two parts). N. Engl. J. Med. 310: 697-703
583 Willmore, L.J., Triggs, W.J. (1984) Effect of phenytoin and corticosteroids on seizures and lipid peroxidation in experimental posttraumatic epilepsy. J. Neurosurg. 60: 467-472
584 Wilson, C.B., Tsai, V., Remington, J.S. (1980) Failure to trigger oxidative metabolic burst by normal macrophages possible mechanism for survival of intracellular pathogens. J. Exp. Med. 151: 328-346
585 Winterbourn, C.C., Williamson, D., Vissers, M.C.M., Carrell, R.W. (1981) Unstable hemoglobin haemolytic crises: contributions of pyrexia and neutrophil oxidants. Br. J. Haematol. 49: 111-116
586 Winterbourn, C.C., Carrell, R.W. (1974) Studies of hemoglobin denaturation and Heinz body formation in the unstable hemoglobins. J. Clin. Invest. 54: 678-689
587 Witschi, H. (1977) Environmental agents altering lung biochemistry. Fed. Proc. 36: 1631-1633
588 Wolf, H.R.D., Seeger, H.W. (1982) Experimental and clinical results in shock lung treatment with vitamin E. In: Lubin, B., Machlin L.J. (eds) Vitamin E: bi-

ochemical, hematological, and clinical aspects. Ann. NY Acad. Sci. 393: 392-409
589 Yamamoto, M., Shima, T., Uozumi, T., Sogabe, T., Yamada, K., Kawasaki, T. (1983) A possible role of lipid peroxidation in cellular damages caused by cerebral ischemia and the protective effect of alfa-tocopherol administration. Stroke 14: 977-982
590 Yip, R., Mohandas, N., Clarik, M.R., Jains, R., Shohet, S.B., Dallman, P.R. (1983) Red cell membrane stiffness in iron deficiency. Blood 62: 99-106
591 Yoshikawa, T., Furukawa, Y., Wakamatso, Y., Kondo, M. (1981) Protective effect of Levamisole on carbon tetrachloride-induced lipid peroxidation and hepatotoxicity. Biomedicine 35: 208-211
592 Yoshikawa, T., Kakimi, Y., Seto, O., Murakami, M., Furukawa, Y., Sugino, S., Kondo, M. (1985) Role of oxygen radicals in endotoxin-induced shock and disseminated intravascular coagulation in rats. In: Fourth international conference on superoxide and superoxide dismutase, Rome, 1-6 September, abstract p 90
593 Yoshikawa, T., Kondo, M. (1982) Role of vitamin E in the prevention of hepatocellular damage: clinical and experimental approach. In: Lubin, B., Machlin, L.J. (eds) Vitamin E: biochemical, hematological, and clinical aspects. Ann. NY Acad. Sci., 393: 198-200
594 Yoshimitsu, K., Kobayashi, Y., Usui, T. (1984) Decreased superoxide dismutase activity of erythrocytes and leukocytes in Fanconi's anemia. Acta Haematol 72: 208-210
595 Young, R.C., Ozols, R.F., Myers, C.E. (1981) The arthracycline antineoplastic drugs. N. Engl. J. Med. 305: 139-153
596 Young, V. R. (1981) Selenium: a case for its essentiality in man. N. Engl. J. Med. 304: 1228-1229
597 Yurkovskaya, T.N., Beregovskaya, N.N., Balenko, N.V., Chernickenko, I.A., Udovichenko, T.V., Bakanova, G.N. (1984) Kinetics of metal-enzymic paramagnetic centres and free radicals in lung carcinogenesis. Ukr Biochim Zh 56: 646-651
598 Younes, M., Siegers, C.P. (1984) Interrelation between lipid peroxidation and other hepatotoxic events. Biochem. Pharmacol 33: 2001-2003, 1984
599 Zs.-Nagy, I., Floyd, R.A. (1984) Electron spin resonance spectroscopic demonstration of the hydroxyl free radical scavenger properties of dimethylaminoethanol in spin trapping experiments confirming the molecular basis for the biological effects of centrophenoxine. Arch. Gerontol. Geriatr. 3: 297-310
600 Zs.-Nagy, I., Floyd, R.A. (1984) Hydroxyl free radical reactions with amino acids and proteins studied by electron spin resonance spectroscopy and spin-trapping. Biochim Biophys. Acta 790: 238-250
601 Zsoldos, T., Tigyi, A., Montskó, T. (1984) Comparative study of the membrane damaging effect of quartz and carbon in the early phase of experimental silicosis. In: Oxygen free radicals and tissue damage, scientific session, Pécs, 10-11 January 1984
602 National Institutes of Health Workshop Report (1983) Free radicals in promotion-a chemical pathology study section workshop. Cancer Res. 43: 5631-5637
603 Are 90% of cancers preventable. Lancet i: 8013, 685-688
604 Conn, H.O. (ed) (1981) International workshop on (+)-cyanidanol-3 in disease of the liver.
605 Anonymous (1985) Metal chelation therapy, oxygen radicals and human disease. Lancet i: 143-145 (editorial)
607 Anonymous (1982) Pregnancy and the arachidonic acid cascade Lancet i: 997-998 (editorial)

608 Anonymous (1981) The senile amyloidoses. Br Med. J. 282: 6267, 846
609 Anonymous (1984) Vitamin A and cancer. Lancet ii: 325-326 (editorial)
610 Anonymous (1976) Vitamin A and cancer prophylax. Br Med. J. 272 (6026): 2-3
611 Anonymous (1980) Vitamin A, retinol, carotene and cancer prevention. BrMed. J. 280: 6246, 957-958
612 Anonymous (1983) Vitamin E reconsidered. Ann of Internal. Med 98: 253-254 (editorial)
613 Anonymous (1979) Workshop on bronchopulmonary dysplasia. J. Pediatr. 95: 815-919

Subject Index

MIX
Papier aus verantwortungsvollen Quellen
Paper from responsible sources
FSC® C105338

If you have any concerns about our products,
you can contact us on
ProductSafety@springernature.com

In case Publisher is established outside the EU,
the EU authorized representative is:
Springer Nature Customer Service Center GmbH
Europaplatz 3, 69115 Heidelberg, Germany

Printed by Libri Plureos GmbH
in Hamburg, Germany